Available soon:

For more information visit our website

www.oup.com/vsi/

Mark Maslin

CLIMATE
CHANGE

A Very Short Introduction
THIRD EDITION

OXFORD
UNIVERSITY PRESS

OXFORD
UNIVERSITY PRESS

Great Clarendon Street, Oxford, OX2 6DP,
United Kingdom

Oxford University Press is a department of the University of Oxford.
It furthers the University's objective of excellence in research, scholarship,
and education by publishing worldwide. Oxford is a registered trade mark of
Oxford University Press in the UK and in certain other countries

© Eco-Climate Limited 2014

The moral rights of the author have been asserted

First edition published as Global Warming: VSI in 2004
Second edition published as Global Warming: VSI in 2009
Third edition published as Climate Change: VSI in 2014

Impression: 1

Published in the United States of America by Oxford University Press
198 Madison Avenue, New York, NY 10016, United States of America

British Library Cataloguing in Publication Data

Data available

Library of Congress Control Number: 2014942174

ISBN 978-0-19-871904-5

Printed in Great Britain by
Ashford Colour Press Ltd, Gosport, Hampshire

*To Chris Pace (1968–2006) and
Nick Shackleton (1937–2006),
who never saw problems, only solutions*

Contents

Acknowledgements

The author would like to thank the following people: Anne, Chris, Johanna, Alexandra, and Abbie for being there, Emma Ma at OUP, all the staff in the UCL, Rezatec Ltd, TippingPoint, and Eden; Miles Irving for excellent illustrations; Richard Betts, Mark Brandon, and Eric Wolff for their insightful and extremely helpful reviews; and all my colleagues in climatology, palaeoclimatology, social science, economics, medicine, engineering, arts, and humanities who continue to strive to understand, predict, and mitigate our influence on climate.

Preface to the third edition

Climate change is one of the few scientific theories that makes us examine the whole basis of modern society. It is a challenge that has politicians arguing, sets nations against each other, queries individual lifestyle choices, and ultimately asks questions about humanity's relationship with the rest of the planet. The latest Intergovernmental Panel on Climate Change (IPCC) report states that the evidence for climate change is unequivocal; with evidence over the last 100 years of a 0.8°Celsius (C) rise in global temperatures and a 22 centimetres (cm) rise in sea level. Depending on how much we control future greenhouse gas (GHG) emissions the global mean surface temperature could rise between 2.8°C and 5.4°C by the end of the 21st century. In addition global sea level could rise by between 52 cm and 98 cm and there will be significant changes in weather patterns with more extreme climate events. This is not the end of the world as envisaged by many environmentalists in the late 1980s and early 1990s, but it does mean a huge increase in misery for billions of people.

Reducing GHG emissions is a major challenge for our global society. This should not be underestimated because despite 30 years of climate change negotiations there has been no deviation in GHG emissions from the business-as-usual pathway. The failure of the international climate negotiation, most notably at Copenhagen in 2009, set back meaningful global cuts in GHG

emissions by at least a decade. Anticipation and hope is building for future negotiations and there are some glimmers of hope. China, now the largest GHG polluter in the World, has started discussing instigating a national carbon-trading scheme. While the USA, which has emitted a third of all the carbon pollution in the atmosphere, has placed the responsibility for regulating carbon dioxide emissions under the Environment Protection Agency, which places it at arm's length to the political wrangling in Washington.

Despite this lack of political agreement there is a strong economic argument for taking action. It is estimated that tackling climate change now would cost between 2–3 per cent of world GDP as opposed to over 20 per cent if we put off action till the middle of the century. Even if the cost–benefits were not so great, the ethical case for paying now to prevent the deaths of tens of millions of people and avoiding a significant increase in human misery must be clear. An international political solution is an imperative, without a post-2015 agreement we are looking at huge increases in global carbon emissions and severe climate change. Any political agreement will have to include developing countries, while protecting their right to develop, as it is a moral imperative that people in the poorest countries should be able to obtain a similar level of health care, education, and life expectancy as those living in the West. Climate change policies and laws based around the international negotiations must also be implemented at both regional and national level to provide multi-levels of governance to ensure these cuts in emissions really do occur. Novel ways of redistributing wealth, globally and well as within nation-states, are needed to lift billions of people out of poverty without huge increases in consumption, resource depletion, and GHG emissions. Support and money is also needed to help developing countries adapt to the climate changes that will inevitably happen.

Climate change, therefore, challenges the very way we organize our society. Not only does it challenge the concept of the

nation-state versus global responsibility, but the short-term vision of our political leaders. Climate change also needs to be seen within the context of the other great challenges of the 21st century: global poverty, population growth, environmental degradation, and global security. To meet these 21st century challenges we must change some of the basic rules of our society, to allow us to adopt a much more global and long-term approach, and in doing so, develop a win-win solution that benefits everyone.

Abbreviations

AABW	Antarctic Bottom Water
AO	Arctic Oscillation
AOGCM	Atmosphere–Ocean General Circulation Model
AOSIS	Alliance of Small Island States
BINGO	Business and Industry Non-Governmental Organization
CCS	carbon capture and storage
CDM	Clean Development Mechanism
CFCs	chlorofluorocarbons
COP	Conference of the Parties
ENGO	Environmental Non-Governmental Organization
ENSO	El Niño–Southern Oscillation
ETS	Emissions Trading Scheme
GCM	General circulation model
GCR	galactic cosmic ray
GHCN	Global Historical Climate Network
GHG	greenhouse gas
GtC	Gigatonnes of Carbon
IPCC	Intergovernmental Panel on Climate Change
JUSSCANNZ	Japan, USA, Switzerland, Canada, Australia, Norway, and New Zealand
MAT	marine air temperature
NADW	North Atlantic Deep Water
NAO	North Atlantic Oscillation
NGO	non-governmental organization
NRC	National Research Council (USA)
OECD	Organization for Economic Cooperation and Development
OPEC	Organization of Petroleum Exporting Countries
PETM	Palaeocene–Eocene Thermal Maximum
ppbv	parts per billion by volume
ppmv	parts per million by volume
SRES	Special Report on Emission Scenarios by the IPCC (2000)
SSS	sea-surface salinity
SST	sea-surface temperature
THC	thermohaline circulation
UNFCCC	United Nations Framework Convention on Climate Change
VBD	vector-borne disease

List of illustrations

© Joel Pitt

List of tables

List of boxes

Chapter 1
What is climate change?

Future climate change is one of the defining challenges of the 21st century, along with poverty alleviation, environmental degradation, and global security. The problem is that 'climate change' is no longer just a scientific concern, but encompasses economics, sociology, geopolitics, national and local politics, law, and health just to name a few. This chapter will examine the role of greenhouse gases (GHGs) in moderating past global climate, why they have been rising since the industrial revolution, and why they are now considered dangerous pollutants. It will examine which countries have produced the most GHGs and how it is changing with rapid development. It will introduce the Intergovernmental Panel on Climate Change (IPCC) and how it regularly collates and assesses the most recent evidence for climate change.

The Earth's natural greenhouse

The temperature of the Earth is determined by the balance between energy from Sun and its loss back into space. Of Earth's incoming solar short-wave radiation (mainly ultraviolet (UV) radiation and visible 'light') nearly all of it passes through the atmosphere without interference (see Figure 1). The only exception is ozone that luckily for us absorbs energy in the

high-energy UV band restricting how much reaches the surface of the Earth as it is very damaging to cells and DNA. About one-third of the solar energy is reflected straight back into space. The remaining energy is absorbed by both the land and ocean. This warms them up, and they then radiate this acquired warmth as long-wave infrared or 'heat' radiation. Atmospheric gases such as water vapour, carbon dioxide (CO_2), methane (CH_4), and nitrous oxide are known as greenhouse gases (GHGs) as they can absorb some of this long-wave radiation, thus warming the atmosphere. This effect has been measured in the atmosphere and can be reproduced time and time again in the laboratory. We need this greenhouse effect because without it, the Earth would be at least 35°Celsius (C) colder, making the average temperature in the tropics about −10°C. Since the industrial revolution we have been burning fossil fuels (oil, coal, natural gas) deposited hundreds of millions years ago, releasing the carbon back into the atmosphere as CO_2 and CH_4, increasing the 'greenhouse effect' and elevating the temperature of the Earth. In effect we are burning fossilized sunlight.

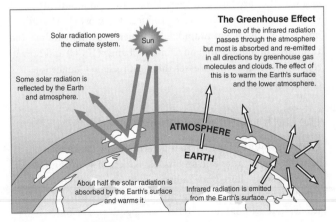

The Greenhouse Effect

Solar radiation powers the climate system.

Some of the infrared radiation passes through the atmosphere but most is absorbed and re-emitted in all directions by greenhouse gas molecules and clouds. The effect of this is to warm the Earth's surface and the lower atmosphere.

Some solar radiation is reflected by the Earth and atmosphere.

ATMOSPHERE

EARTH

About half the solar radiation is absorbed by the Earth's surface and warms it.

Infrared radiation is emitted from the Earth's surface.

1. The greenhouse effect

Past climate

Climate change in the geological past has been reconstructed using a number of key archives, including marine and lake sediments, ice cores, cave deposits, and tree rings. These various records reveal that over the last 50 million years the Earth's climate has been cooling down, moving from the so-called 'greenhouse world' of the Eocene, with warm and gentle conditions, through to the cooler and more dynamic 'ice house world' of today. It may seem odd that in geological terms our planet is extremely cold, while this whole book is concerned with our rapid warming of the planet. This is because the very fact that there are huge ice sheets on both Antarctica and Greenland, and nearly permanent sea ice in the Arctic Ocean, makes the global climate very sensitive to changes in GHGs.

The long-term global cooling of the Earth kicked off with glaciation of Antarctica about 35 million years ago and then the great Northern Hemisphere ice ages, which began 2.5 million years ago. Since the beginning of the great northern ice ages, the global climate has cycled from conditions that were similar or even slightly warmer than today, to full ice ages, which caused ice sheets over 3 kilometres (km) thick to form over much of North America and Europe. Between 2.5 and 1 million years ago, these glacial–interglacial cycles occurred every 41,000 years, and since 1 million years ago they have occurred every 100,000 years. These great ice-age cycles are driven primarily by changes in the Earth's orbit with respect to the Sun. In fact, the world has spent over 80 per cent of the last 2.5 million years in conditions colder than the present. Our present interglacial, the Holocene Period, started about 10,000 years ago, and is an example of the rare warm conditions that occur between each ice age. The Holocene began with the rapid and dramatic end of the last ice age: in less than 4,000 years global temperatures increased by 6°C, relative sea level rose by 120 metres (m), atmospheric CO_2 increased

by one-third, and atmospheric CH_4 doubled. Still, this is much slower than changes we are seeing today. James Lovelock in his book *The Ages of Gaia* suggests that interglacials, like the Holocene, are the fevered state of our planet, which clearly over the last 2.5 million years prefers a colder average global temperature. Lovelock sees global warming as humanity just adding to the fever. These large scale past changes in global climate are discussed in more detail in my other book *Climate: A Very Short Introduction*.

Past variations in carbon dioxide

One of the ways in which we know that atmospheric CO_2 is important in controlling global climate is through this study of past climate. Evidence for these past variations in GHGs and temperature come from ice cores drilled in both Antarctica and Greenland. As snow falls, it is light and fluffy and contains a lot of air. When this snow is slowly compacted to form ice, some of this air is trapped. By extracting these air bubbles trapped in the ancient ice, scientists can measure the percentage of GHGs that were present in the past atmosphere. Scientists have drilled over two miles down into both the Greenland and Antarctic ice sheets, which has enabled them to reconstruct the amount of GHGs that have been generated in the atmosphere over the last half a million years. By examining the oxygen and hydrogen isotopes in the frozen water that make up the ice core, it is possible to estimate the air temperature above the ice sheet when the water first froze. The results are striking, as GHGs such as atmospheric CO_2 and CH_4 co-vary with temperatures over the last 800,000 years (see Figure 2). The cyclic changes in climate from glacial to interglacial periods can be seen both in temperatures and the GHG content of the atmosphere. This strongly supports the idea that GHGs in the atmosphere and global temperature are closely linked, that is, when CO_2 and CH_4 increase, the temperature is found to increase and vice versa.

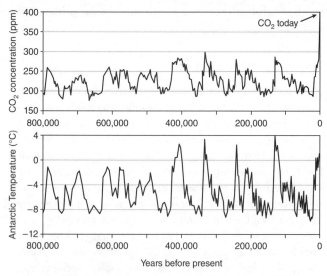

2. **Greenhouse gases and temperature for the last eight glacial cycles recorded in ice cores**

Early farmers

The very high-resolution ice-core evidence from Greenland and the continental margins of Antartica also show that GHGs in the atmosphere rose a small amount before the industrial revolution in the 1700s. Bill Ruddiman, Professor of Palaeoclimatology at the University of Virginia, suggested that early agriculturalists caused a reversal in natural declines of atmospheric CO_2 starting about 7,000 years ago and atmospheric CH_4 starting about 5,000 years ago. This idea has caused a huge amount of controversy, but like all good theories it has been tested again and again, and no one has yet been able to disprove it. So essentially it says that early human interactions with our environment increased atmospheric GHGs just enough that even prior to the industrial revolution there was enough influence to delay the onset of the next ice age,

which would otherwise have started gently to occur anytime in the next 1,000 years.

The industrial revolution

There is clear evidence that levels of atmospheric CO_2 have been rising ever since the beginning of the industrial revolution. The first measurements of CO_2 concentrations in the atmosphere started in 1958 at an altitude of about 4,000 m, on the summit of Mauna Loa Mountain in Hawaii. The measurements were made here to be remote from local sources of pollution. The record clearly shows that atmospheric concentrations of CO_2 have increased every single year since 1958. The mean concentration of approximately 316 parts per million by volume (ppmv) in 1958 has risen to over 400 ppmv now (see Figure 3). The annual variations in the Mauna Loa observatory are mostly due to CO_2 uptake by growing plants. The uptake is highest in the Northern Hemisphere springtime; hence every spring there is a drop in atmospheric CO_2, which

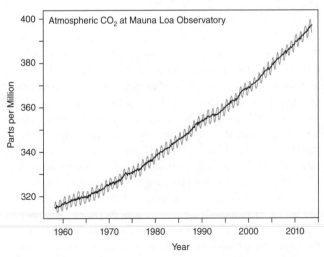

3. **Mauna Loa observatory atmospheric carbon dioxide measurements**

unfortunately does nothing to change the overall trend towards ever higher values.

This CO_2 data from the Mauna Loa observatory can be combined with the detailed work on ice cores to produce a complete record of atmospheric CO_2 since the beginning of the industrial revolution. This shows that atmospheric CO_2 has increased from a pre-industrial concentration of about 280 ppmv to over 400 ppmv at present, representing an increase of over 40 per cent. To put this increase into context, ice-core evidence shows that over the last 800,000 years the natural change in atmospheric CO_2 has been between 180 and 300 ppmv. The variation between warm and cold periods is about 80 ppmv—less than the CO_2 pollution that we have put into the atmosphere over the last 100 years. It demonstrates that the level of pollution that we have already caused in one century is comparable to the natural variations which took thousands of years.

Who produces the pollution?

The United Nations Framework Convention on Climate Change (UNFCCC) was created to produce the first international agreement on reducing global GHG emissions. However, this task is not as simple as it first appears, as CO_2 emissions are not evenly produced by countries. According to the Intergovernmental Panel on Climate Change (see Box 1) the first major source of CO_2 is the burning of fossil fuels, since four-fifths of global CO_2 emissions comes from energy production, industrial processes, and transport. These are not evenly distributed around the world because of the unequal distribution of industry and wealth; North America, Europe, and Asia emit over 90 per cent of the global, industrially produced CO_2 (see Figure 4). Moreover, historically the developed nations have emitted much more than less developed countries.

The second major source, accounting for one-fifth of global CO_2 emissions, is as a result of land-use changes. These emissions

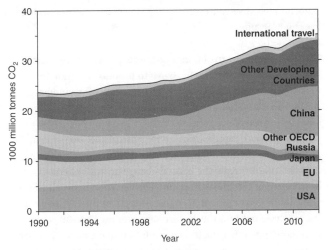

4. **Historic carbon dioxide emissions by region**

come primarily from the cutting down of forests for the purposes of agriculture, urbanization, or roads. When large areas of rainforests are cut down, the land often turns into less productive grassland with considerably reduced capacity for storing CO_2. Here the pattern of CO_2 emissions is different, with South America, Asia, and Africa being responsible for over 90 per cent of present-day land-use change emissions. This raises important ethical questions because it is difficult to tell these countries to stop deforestation when historically this had already occurred in much of North America and Europe before the beginning of the 20th century. In terms of the amount of CO_2 released, industrial processes still significantly outweigh land-use changes.

So who are the bad guys in causing this increase in atmospheric CO_2? Of course, it is the developed countries who historically have emitted most of the anthropogenic (man-made) GHGs, as they have been emitting them since the start of the industrial revolution in the latter half of the 18th century. Though this

historic carbon burden is important according to International Energy Authority projections, it is rapidly changing. Between 2015 and 2044 the world will put half a trillion tonnes of CO_2 into the atmosphere, which is the same amount that was emitted between 1750 and 2015. This is because rapidly developing countries such as China, India, South Africa, Brazil, etc., are increasing their emissions of GHGs at a huge rate—economic development being closely associated with energy production. For example, in 2007, China become the biggest emitter of CO_2 in the world, overtaking the USA. However, when considered per capita, the Chinese emissions are four times lower than those of the USA, who are top of the per-capita list.

The half trillion tonnes of carbon which have been put into the atmosphere since the industrial revolution represent only half our total emission. It seems that 50 per cent of our emission have been absorbed by the Earth, with 25 per cent going into the oceans and 25 per cent going into the land biosphere. However, scientists are concerned as this removal of our pollution is unlikely to continue fully in the future. This is because as global temperatures rise the oceans will warm and will be able to hold less CO_2. As we continue to deforest and convert land for farming and urbanization there will be less vegetation to absorb CO_2, again reducing the uptake of our carbon pollution.

Box 1 What is the IPCC?

The Intergovernmental Panel on Climate Change (IPCC) was established in 1988 jointly by the United Nations Environmental Panel and the World Meteorological Organization to address concerns about the possibility of global warming. The purpose of the IPCC is the continued assessment of the state of knowledge on the various aspects of climate change, including scientific, environmental, and socioeconomic impacts and response strategies. The IPCC does not undertake independent scientific

Box 1 Continued

research, rather it brings together all key research published in the world and produces a consensus. There have been five main IPCC Reports, in 1990, 1996, 2001, 2007, and 2013/14, and many individual specalized reports on such subjects as carbon emission scenarios, alternative energy sources, and extreme weather events.

The IPCC is, thus, recognized as the most authoritative scientific and technical voice on climate change, and its assessments have had a profound influence on the negotiators of the United Nations Framework Convention on Climate Change (UNFCCC). The IPCC is organized into three working groups plus a task force to calculate the amount of GHGs produced by each country. Each of these four bodies has two co-chairmen (one from a developed and one from a developing country) and a technical support unit. Working Group I assesses the scientific aspects of the climate system and climate change; Working Group II addresses the vulnerability of human and natural systems to climate change, the negative and positive consequences of climate change, and options for adapting to them; and Working Group III assesses options for limiting GHGs emissions and otherwise mitigating climate change, as well as economic issues. Hence the IPCC also provides governments with scientific, technical, and socioeconomic information relevant to evaluating the risks and to developing a response to global climate change. The latest reports from these three working groups were published in 2013 and 2014, with approximately 500 experts, from some 120 countries, directly involved in drafting, revising, and finalizing the IPCC reports, as well as over 2,000 experts participating in the review process. The IPCC authors are always nominated by governments and international organizations, including non-governmental organizations. These reports are essential reading for anyone interested in global warming and are listed in the Further reading section at the end of the book. In 2008, the IPCC was jointly awarded, with Al Gore, the Nobel Peace Prize, to acknowledge all the work the IPCC has done over the past 20 years.

Summary

There is clear evidence that GHGs concentrations in the atmosphere have been rising since the industrial revolution in the 18th century. Atmospheric concentrations of both CO_2 and CH_4 are higher than anytime within at least the last million years. Within 100 years we have put more carbon into the atmosphere than the amount of carbon that was emitted between the long natural glacial–interglacial cycles, which took thousands of years. The current scientific consensus is that these recent changes in GHG concentrations in the atmosphere have already caused the global temperatures to increase. Since 1880, the global average surface temperature has increased by 0.85°C. This warming has been accompanied by a significant warming of the ocean, a rise in sea level of 20 centimetres (cm), a 40 per cent decline in Arctic sea ice, and an increase in number of extreme weather events. As we emit more and more carbon into the atmosphere the effects in climate change will increasingly threaten and challenge human society. The science, politics, and potential solutions to climate change are examined in the rest of this book. In Chapter 2 the recognition of climate change as a global pollution problem is examined. Chapters 3 and 4 discuss the current scientific evidence for climate change and how scientists are modelling the future to assess how global carbon emissions will alter our climate. Chapters 5 and 6 examine the impacts of these future climate changes and the possibility that there may be hidden surprises within the climate system that may exacerbate climate change. Chapters 7 and 8 investigate the political aspects of climate change, and potential political and technological solutions. Finally, Chapter 9 provides multiple views of the future dependent on how we decide to tackle carbon emissions and the impacts of climate change.

Chapter 2
The climate change debate

Historical background

Scientists are predicting that continuing on our current carbon emissions pathway we could warm the planet by between 2.8 and 5.6°C in the next 85 years, which economists suggest could cost us as much as 20 per cent of world gross domestic product (GDP) to deal with. In the face of such a threat, it is essential to understand the history of climate change and the evidence that supports it. The essential science of climate change was carried out 50 years ago under the perceived necessity of geosciences during the Cold War, but it was not taken seriously until the late 1980s. Since then climate change has emerged as one of the biggest scientific and political problems facing humanity.

It is now over 100 years since 'global warming' was officially discovered. The pioneering work in 1896 by the Swedish scientist Svante Arrhenius, and the subsequent independent confirmation by Thomas Chamberlin, calculated that human activity could substantially warm the Earth by adding carbon dioxide (CO_2) to the atmosphere. This conclusion was the by-product of other research, the central aim of which was to provide a theory suggesting that decreased CO_2 was a major cause of the great ice ages. The theory still stands today but had to wait until 1987 for the Antarctic Vostok ice-core results to confirm the pivotal role of

atmospheric CO_2 in controlling past global climate. However, no one else took up the research topic, so both Arrhenius and Chamberlin turned to other challenges. This was because scientists at that time felt there were so many other influences on global climate, from sunspots to ocean circulation, that minor human influences were thought insignificant in comparison to the mighty forces of astronomy and geology. This idea was reinforced by research during the 1940s, which developed the theory that changes in the orbit of the Earth around the Sun controlled the waxing and waning of the great ice ages. A second line of argument was that because there is 50 times more CO_2 in the oceans than in the atmosphere, 'The sea acts as a vast equalizer': in other words, the ocean would mop up our pollution.

This dismissive view took its first blow when in the 1940s there was a significant improvement in infrared spectroscopy, the technique used to measure long-wave radiation. Up until then, experiments had shown that CO_2 did block the transmission of infrared 'long-wave' radiation of the sort given off by the Earth. However, the experiments showed there was very little change in this interception if the amount of CO_2 was doubled or halved. This meant that even small amounts of CO_2 could block radiation so thoroughly that adding more gas made very little difference. Moreover, water vapour, which is much more abundant than CO_2, was found to block radiation in the same way and, therefore, was thought to be more important.

The Second World War saw a massive improvement in technology and the old measurements of CO_2 radiation interception were revisited. In the original experiments sea-level pressure was used, but it was found that at the rarefied upper atmosphere pressures the general absorption did not occur and, therefore, radiation was able to pass through the upper atmosphere and into space. This proved that increasing the amount of CO_2 did result in absorption of more radiation. Moreover, it was found that water vapour absorbed other types of radiation rather than CO_2, and to

compound it all, it was also discovered that the stratosphere, the upper atmosphere, was bone dry. This work was brought together in 1955 by the calculations of Gilbert Plass, who concluded that adding more CO_2 to the atmosphere would intercept more infrared radiation, preventing it being lost to space and thus warming the planet.

This still left the argument that the oceans would soak up the extra anthropogenically produced CO_2. The first new evidence came in the 1950s and showed that the average lifetime of a CO_2 molecule in the atmosphere before it dissolved in the sea was about ten years. As the ocean overturning takes several hundreds of years, it was assumed the extra CO_2 would be safely locked in the oceans. But Roger Revelle, director of Scripps Institute of Oceanography in California, realized that it was necessary not only to know that a CO_2 molecule was absorbed after ten years but to ask what happened to it after that. Did it stay there or diffuse back into the atmosphere? How much extra CO_2 could the oceans hold? Revelle's calculations showed that the complexities of surface ocean chemistry are such that it returns much of the CO_2 that it absorbs. This was a great revelation, and showed that because of the peculiarities of ocean chemistry, the oceans would not be the complete sink for anthropogenic CO_2 that was first thought. This principle still holds true, although the exact amount of anthropogenic CO_2 taken up per year by the oceans is still in debate, it is about one-quarter of the annual total anthropogenic production.

Charles Keeling, who was hired by Roger Revelle, produced the next important step forward in the global warming debate. In the late 1950s and early 1960s, Keeling used the most modern technology available to measure the concentration of atmospheric CO_2 in Antarctica and Mauna Loa. The resulting Keeling CO_2 curves have continued to climb ominously each year since the first measurement in 1958 and have become one of the major icons of global warming (Figure 3).

Cold War science

Spencer Weart, the director of the Center of History of Physics at the American Institute of Physics, contends that all the scientific facts about enhanced atmospheric CO_2 and potential global warming were assembled by the late 1950s to early 1960s. He argues that it was only due to the physical geosciences being favoured financially in the Cold War environment that so much of the fundamental work on global warming was completed. Gilbert Plass published an article in 1959 in *Scientific American* declaring that the world's temperature would rise by 3°C by the end of the century. The magazine editors published an accompanying photograph of coal smoke belching from factories and the caption read, 'Man upsets the balance of natural processes by adding billions of tons of carbon dioxide to the atmosphere each year.' This resembles thousands of magazine articles, television news items, and documentaries that we have all seen since the late 1980s. So why was there a delay between the science of global warming being accepted and in place in the late 1950s and early 1960s and the sudden realization by those outside the scientific community of the true threat of global warming during the late 1980s?

Why the delay in recognizing climate change?

The key reasons for the delay in recognizing the climate change were, first, the power of the global mean temperature (GMT) data set and, second, the need for the emergence of global environmental awareness. The GMT data set is calculated using the land-air and sea-surface temperatures. From 1940 until the mid-1970s, the global temperature curve seems to have had a general downward trend. This provoked many scientists to discuss whether the Earth was entering the next great ice age. This fear developed in part because of increased awareness in the 1970s of how variable global climate had been in the past. The emerging subject of palaeoceanography (study of past oceans) demonstrated

from deep-sea sediments that there were at least 32 glacial–interglacial (cold–warm) cycles in the last 2.5 million years, not 4 as had been previously assumed. The time resolution of these studies was low, so that there was no possibility of estimating how quickly the ice ages came and went, only how regularly. It led many scientists and the media to ignore the scientific revelations of the 1950s and 1960s in favour of global cooling. It was not until the early 1980s, when the global annual mean temperature curve started to increase, that the global cooling scenario was questioned. By the late 1980s, the global annual mean temperature curve rose so steeply that all the dormant evidence from the late 1950s and 1960s was given prominence and the global warming theory was in full swing.

It seems that the eventual recognition of the climate change issue was driven by the upturn in the global annual mean temperature data set. The latest IPCC 2013 Science Report has reviewed and synthesized a wide range of data sets and shows that, essentially, the trend in global temperature first recognized in the late 1980s is essentially correct, and that this warming trend has continued unstopped until the present day (see Figure 5). In fact, we know that 1998 and 2010 were globally the warmest years on record. The temperatures for these two years are so close that scientists are divided on which is the warmest. However, 1998 was an El Niño year, which we know adds up to 1°C on the average global temperatures. So the warmest 'normal' year on record is 2010. If the global temperature records are analysed in decadal blocks it is even clearer that the last three decades are all significantly warmer than the previous ones (Figure 5).

The upturn in the global annual mean temperature data was not the sole reason for the appearance of the global warming issue. In the late 1970s and 1980s, there were significant advances in global climate modelling and a marked improvement in our understanding of past climates. Developments in general

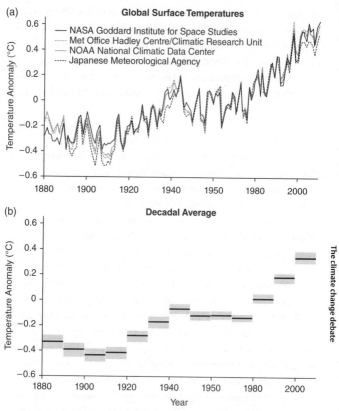

5. Variation of the Earth's surface temperature over the last 150 years

circulation models (GCMs) during this period included taking
into account the role of particles and clouds in affecting the
global climate. Despite the cooling effect thought to be associated
with particle pollution, the new ocean–atmosphere-coupled GCM
tools emerged with revised and higher estimates of the warming
that would be associated with a doubling of CO_2 in the
atmosphere. By the 1980s, scientific concern had emerged about

methane (CH_4) and other non-CO_2 GHGs as well as the role of the oceans as a carrier of heat. GCMs continued to improve, and the numbers of scientific teams working on such models increased over the 1980s and the 1990s. In 1992, a first overall comparison of results from 14 GCMs was undertaken; the results were all in rough overall agreement, confirming the prediction of global warming.

In terms of the study of palaeoclimate, during the 1980s there was also an intense drive to understand how and why past climate changed. Major advances were made in obtaining high-resolution past climate records from deep-sea sediments and ice cores. It was, thus, realized that glacial periods, or ice ages, take tens of thousands of years to occur, primarily because ice sheets are very slow to build up and are naturally unstable. In contrast, the transition to a warmer period, or interglacial, such as the present, is geologically very quick, in the order of a couple of thousand years. This is because once the ice sheets start to melt, there are a number of positive feedbacks that accelerate the process, such as sea-level rise, which can undercut and destroy large ice sheets. The realization occurred in the palaeoclimate community that global warming is much easier and more rapid than cooling. It also put to rest the myth of the next impending ice age. We now know that glacial–interglacial periods of the last 2.5 million years were caused by the climate system being influenced by the changes in the orbit of the Earth around the Sun. It is, thus, possible to predict when the next glacial period will begin, if there were no anthropogenic effects involved, which should be sometime in the next 1,000 years. According to the model predictions by Professor David Archer (University of Chicago) and Professor Andrey Ganopolski (Potsdam University) current levels of GHG emissions have already delayed the next glacial period by 40,000 years, if we did everything possible to reduce future emissions then the resultant carbon pulse would delay the next glacial by 120,000 years. If emissions continue to rise at a 'business as usual' rate we could delay the next glacial by over half a million years.

The rise of the environmental social movement

The next change that occurred during the 1980s was a massive grass-roots expansion in the environmental movement, particularly in the USA, Canada, and the UK, partly as a backlash against the right-wing governments of the 1980s and the expansion of the consumer economy, and partly because of the increasing number of environment-related stories in the media. This heralded a new era of global environmental awareness and transnational non-governmental organizations (NGOs). The roots of this growing environmental awareness can be traced back to a number of key markers: these include the publication of Rachel Carson's *Silent Spring* in 1962; the image of Earth seen from the Moon in 1969; the Club of Rome's 1972 report on *Limits to Growth*; the Three Mile Island nuclear reactor accident in 1979; the nuclear accident at Chernobyl in 1986; and the Exxon Valdez oil spillage in 1989. But these environmental problems were all regional in effect, limited geographically to the specific areas in which they occurred.

It was the discovery in 1985 by the British Antarctic Survey of depletion of ozone over Antarctica, which demonstrated the global connectivity of our environment. The ozone 'hole' also had a tangible international cause, the use of chlorofluorocarbons (CFCs), which led to a whole new area of politics: the international management of the environment. There followed a set of key agreements: the 1985 Vienna Convention for the Protection of the Ozone Layer; the 1987 Montreal Protocol on Substances that Deplete the Ozone Layer; and the 1990 London and 1992 Copenhagen Adjustments and Amendments to the Protocol. These have been held up as examples of successful environmental diplomacy. Climate change has had a slower development in international politics and far less has been achieved in terms of regulation and implementation, primarily because it is so much more complicated and the fact that fossil fuels are currently central to all industrial development.

Climate change and the media

The other reason for the emergence of climate change as a major global issue was the intense media interest throughout the late 1980s and 1990s. This is because climate change is perfect for the media: a dramatic story about the end of the world as we know it, with significant controversy about whether it was even true. Anabela Carvalho, now at the University of Minho (Braga, Portugal), undertook a fascinating study of the British quality (broadsheet) press coverage of the global warming issue between 1985 and 1997. She concentrated particularly on the *Guardian* and *The Times*, and found throughout this period that they promoted very different worldviews. Interestingly, despite their differing views, the number of articles published per year by the British quality papers followed a similar pattern and peaked when key IPCC reports were published or international conferences on climate change were held (see Figure 6). But it is the nature of these articles that shows how the global warming debate was constructed in the media. From the late 1980s, *The Times*, which published most articles on global warming in 1989, 1990, and 1992, cast doubt on the claims of climate change. There was a recurrent attempt to promote mistrust in science, through strategies of generalization, of disagreement within the scientific community, and, most importantly, discrediting scientists and scientific institutions. A very similar viewpoint was taken by the majority of the American media throughout much of the 1990s. In fact, it has been claimed that this approach in the American media has led to a barrier between scientists and the public in the USA.

In the UK, the *Guardian* newspaper took the opposite approach to that of *The Times*. Although the *Guardian* gave space to the technical side of the debate, it soon started to discuss scientific claims in the wider context. As scientific uncertainty regarding the enhanced greenhouse effect decreased during the 1990s, the *Guardian* coherently advanced a strategy of building confidence in science, with an emphasis on consensus as a means of

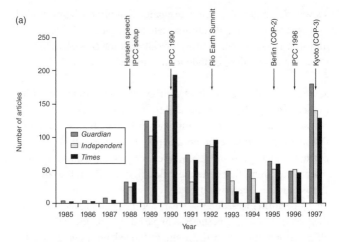

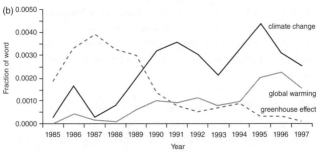

6. Newspaper coverage of global warming 1985–97

enhancing the reliability of knowledge. This was because the
Guardian understood and promoted one of the fundamental
bases of science, which is that a theory, such as global warming,
can only be accepted or rejected by the weight of evidence. So, as
evidence from many different areas of science continues to
support the theory of global warming, so correspondingly our
confidence in the theory should increase. Far from painting
science as 'pure' or 'correct', instead the *Guardian* politicized it to
demonstrate the bias that is inherent in all science. This clearly

21

showed that many of the climate change claims were being eroded by lobbying pressure, mainly associated with the fossil-fuel industry. This politicizing of science allowed the *Guardian* to strengthen its readers' confidence in science.

Moreover, the *Guardian* clearly conveyed the uncertainties within the science of global warming, and the paper's editors were, and still are, in favour of the precautionary principle. It was through this media filter that scientists attempted to advance their particular global warming view, by either making claims for more research or promoting certain political options. From the late 1980s onwards, scientists became very adept at staging their media performances, and it is clear that the general acceptance of the global warming hypothesis is in part due to their continued efforts to communicate their findings. Indeed, both the sceptical and the supportive stances of *The Times* and the *Guardian*, respectively, so legitimized the debate over global warming that the public became aware that this was not an overnight news story but something that was to become part of the very fabric of our society.

It seems that the media has also influenced our use of words. From 1988 onwards, the use of the phrases 'global warming' and 'climate change' gained support, while 'greenhouse effect' lost its appeal and by 1997 was rarely mentioned. Since then climate change has become the dominant word, with the realization that changes in precipitation, sea level, and extreme weather events are for humanity more important than an average raise in global temperatures.

As mentioned, in the USA the media coverage has been different. First, until recently there has been no pro-climate change media coverage equivalent to that delivered by the *Guardian*. Second, climate change sceptics have been very strong on using the media in the USA. For example, McInytre and McKitrick in 2003 attacked the GMT 'hockey stick' (Figure 7) by raising questions

about the quality of data and accuracy of methods used to estimate trends in GMT. This debate has taken place between experts but in an unusually public manner. For example, one US scientist, Michael Mann, who has widely published on GMT trends, has been at the centre of the criticism and active in responding to it. Despite Mann's strong rebuttal and the weight of scientific evidence that brings into doubt the validity of the critique, both the media and US and British politicians have continued to bring attention to the questions raised by McInytre and McKitrick long after they have been discredited.

There are two possible explanations for this extraordinarily media-facilitated public scientific debate. First, climate sceptics and industrial lobby groups who do not want to see political action to address climate change are using this debate about methods and scientific uncertainty as a convenient hook on which to hang their case for delay. The GMT curve over the last millennium is a particularly important target for such criticism, owing to its emblematic role in the policy debate. It is also clear that a huge amount of money has been pumped in by these anti-climate change lobby groups and for them any hesitation or delay in action is a victory for these groups. Second, the media's ethical commitment to balanced reporting may unwittingly provide unwarranted attention to critical views, even if they are marginal and outside the realm of what is normally considered 'good' science. In the UK, the BBC has come under increasing criticism for trying to provide this balance. An excellent report by Professor Steve Jones (University College London) in 2011, commissioned by the BBC, pointed out that they were skewing whole scientific debates by setting up one-to-one discussions on radio and TV. In many areas of science, thousands of scientists may agree on a particular point of view or principle yet they will be given only the same amount of time as a few mavericks. It has got even worse recently at the BBC as they are now pitting scientists against political commentators such as Lord Lawson who feels qualified to debunk any scientific evidence he does not agree with.

Add to this the greater ease of communication, from conventional media, such as newspapers, radio, and television, to more informal blogs, tweets, etc. Normal private debate among scientists and experts can easily be shifted into the public arena and anyone, what ever their level of expertise, can voice an opinion and feel it is as valid as that of experts who have dedicated their whole lives to studying areas of science. Overall, this contributes to a public impression that the science of climate change is 'contested', despite what many would argue is an overwhelmingly scientific case that climate change is occurring and human activity is a main driver of this change.

In the USA, other forms of media that rely on visual information such as film, television, and the internet have perhaps been more powerful than newspaper coverage. Researchers have studied the effects of the 2004 Hollywood blockbuster film *The Day After Tomorrow*. With a huge viewing public (estimated at 21 million people in the USA alone), *The Day After Tomorrow* was a commercial success and also appears to have helped to promote climate change from an obscure scientific issue to one of popular public concern. In addition, the media coverage in glossy magazines, which started initially in *Time Magazine* in April 2006, and *Vanity Fair* in May 2006, has begun to convey a greater sense of urgency about climate change. Finally, widespread media coverage of emblematic impacts of climate change has also been stimulated by the international release of the documentary film *An Inconvenient Truth* by Al Gore; the environmental message of James Cameron's film *Avatar* in 2009; and by a growing number of television documentaries on climate change, such as *60 Minutes*, ABC News, and HBO documentaries. This visual approach in the USA has recently culminated in Showtime's nine-part documentary television series *Years of Living Dangerously*, focusing on the human stories of climate change that premiered on 13 April 2014. It has James Cameron, Jerry Weintraub, and Arnold Schwarzenegger as executive producers, and celebrity presenters include Harrison Ford, Matt Damon,

Jessica Alba, Don Cheadle, America Ferrera, Arnold Schwarzenegger, Lesley Stahl, Mark Bittman, Ian Somerhalder, Olivia Munn, and Michael C. Hall. This rise in 'visual' media coverage suggests that in the last decade the communication of climate change has gone mainstream and the emphasis is changing from promoting fear to promoting solutions.

The economists wade in

Economists have been involved with studying climate change from the very beginning of the IPCC process. There are two landmark publications that have had very different effects on the climate change debate. First, there was the publication of the controversial book *The Skeptical Environmentalist* by Bjørn Lomborg in English in 2001. In this and subsequent books, he argues that the cost of cutting global GHG emissions is extremely high and that those who suffer most from the effects of climate change are those in the poorest countries. He argues that a better use of this money would be poverty alleviation and rapid development of the Third World. As he suggests, if you are starving, you are not worried about the state of the planet for your children, you are worrying whether to have children at all. This view, as you can imagine, was highly controversial. The second major landmark was the publication of the UK government-commissioned Stern Report on *The Economics of Climate Change* in 2006 (the Cambridge University Press version first published in 2007). The report was led by Sir Nicholas Stern, then the adviser to the UK government on the economics of climate change and development reporting to the prime minister (then Tony Blair). The report states that if we do nothing, then the impacts of climate change could cost between 5 per cent and 20 per cent of world GDP every year. That means the whole world loses one-fifth of what it earns to address the impacts (discussed in Chapter 5). This of course puts climate change impacts on a completely different economic scale than was envisaged by Lomborg. But the Stern Report does present some good news because if we do everything we can to

reduce global GHG emissions and ensure we adapt to the coming effects of climate change, this will cost us only 1 per cent of world GDP every year. The Stern Report has been criticized on specifics by other economists—for example, does it use the right inherent discount rate? This is the rate economists use to take into account that consumption inherently has a lower value in the future than in the present. In other words, future consumption should be discounted simply because it takes place in the future and people generally prefer the present to the future. For example, William Nordhaus used inherent discount rates of up to 3 per cent, which means an environmental benefit occurring 25 years in the future is worth about half as much as the same benefit today. The Stern Report has also been criticized for being overly optimistic about the costs of adapting to a low-carbon world. In June 2008, Sir Nicholas Stern did revise his estimated costs up to 2 per cent of world GDP. Nevertheless, the Stern Report sent seismic waves around the world. It was as if people said to themselves, 'if the economists are worried about the cost of climate change, it must be real'.

Climategate

In November 2009 just before the Copenhagen Climate Conference there was an illegal release, due to hacking, of thousands of emails and other documents from the University of East Anglia's (UEA) Climatic Research Unit (CRU). This sparked the infamous 'Climategate scandal'. Allegations were made that the emails revealed misconduct within the climate science community including withholding scientific information, preventing papers being published, deleting raw data, and manipulating data to make the case for climate change to appear stronger than it is. Three independent inquiries since then have concluded that there was no evidence of scientific malpractice. But what it does show is the power of the media and how a story is as good as the coverage it gets, not whether it is true or not. So for the first few days after the emails were released and key phrases

and exchanges were put online most science journalists ignored the story. As far as they were concerned this is exactly how scientists behave and it was not telling them anything new. Then the chief editors got involved as this was a story and clearly they thought the science journalists had gone native, so the story was passed to the political journalists just in time for the Copenhagen conference. But of course what these journalists and editors ignored was that at least two other highly respected groups at National Oceanic and Atmospheric Administration (NOAA) and National Aeronautics and Space Administration (NASA) had used different raw data sets and different statistical approaches and had published the very same conclusions as the UEA group. This was further supported in 2012 when Professor Richard Muller, a physicist and previously a climate change sceptic, and his Berkeley group published their collated global temperature records for the last 200 years, when he publicly announced he had changed his mind and that climate change was occurring due to human activity. This demonstrates how non-science journalists appear to have no concept of the 'weight of evidence', discussed in Chapter 3.

Central to the criticism that the UEA group and by extension other climate scientists had changed the raw data was the potentially misleading short-hand terminology used by scientist such as 'correct', 'trick', 'tweak', 'manipulate', 'a line', 'correlate'. Some raw data does need to be processed so it can be compared with other data, particularly if you are trying to make long records of temperature when the methods used to measure temperature have changed. The clearest example of this is the measurement of sea temperature, which over the last 150 years has varied from a bucket of seawater hoisted on deck to direct measurements on seawater taken into a ships' engine. If scientists just stuck all this raw data together it would of course be inaccurate. Moreover, in this case because the earlier sea-surface temperature measurements are too cold without correction it would make global warming in the ocean appear much greater than it really is. So the constant checking and correcting of data is extremely

important in all parts of science. But the most important questions are: can the results be reproduced? And is there the weight of evidence from many research groups to show the same changes? This is why, after over 30 years of intensive research into climate change, most scientists have a very high level of confidence that climate change is happening and that it is due human activity.

Summary

So a combination of several factors—(1) the science of climate change essentially carried out by the mid-1960s; (2) the 'hockey stick' upturn in the global temperature data set, which was first observed at the end of the 1980s; (3) our increased knowledge in the 1980s of how past climate has reacted to changes in atmospheric CO_2; (4) our greater ability in the late 1970s and 1980s to model future changes in climate with supercomputers; (5) the emergence of global environmental awareness in the late 1980s; (6) the media's savage engagement in the confrontational nature of the debate; and (7) politicians, medics, and economists taking the threat of climate change seriously since the late 1990s—has led to recognition and acceptance that climate change is occurring and that it is anthropogenic.

Chapter 3
Evidence for climate change

Weight of evidence

Science is not a belief system. For example you cannot decide that you believe in antibiotics (as they may save your life) or that heavy metal tubes with wings can fly (because you want to go on holiday), and yet at the same time deny that smoking causes cancer, or that HIV causes AIDS, or that GHGs cause climate change. Science is based on a rational methodology that moves forward by using detailed observation and experimentation to constantly test ideas and theories. It is the very foundation of our global society. If we are to understand climate change, we must understand how science works and the principle of the 'weight of evidence', which prompts the continual need to compile new data and undertake new experiments to test our ideas and theories regarding climate. Over the last 30 years the theory of climate change must have been one of the most comprehensively tested ideas in science. There are six main areas of evidence that should be considered. First, as described above we have tracked the rise in GHGs in the atmosphere and understand their role in past climate variations. Second, we know from laboratory and atmospheric measurements that these gases do indeed absorb heat when they are present in the atmosphere. Table 1 summarizes the latest understanding of the main GHGs. Third, we have tracked significant changes in global temperatures and sea level

Table 1. Main greenhouse gases and their comparative ability to warm the atmosphere

Greenhouse gas	Chemical formula	Lifetime (years)	Pre-industrial levels	2011 levels	Human source	Global warming potential	
						20 years	100 years
Carbon dioxide	CO_2		278 ppmv	391 ppmv (40% increase)	Fossil-fuel combustion Land-use changes Cement production	1	1
Methane	CH_4	12.4	700 ppbv	1803 ppbv (250% increase)	Fossil-fuels Rice paddies Waste dumps Livestock	84	28
Nitrous oxide	N_2O	121	275 ppbv	324 ppbv (18% increase)	Fertilizer Industrial processes Fossil-fuel combustion	264	265

CFC-12	CCl_2F_2	100	Not naturally occurring	0.528 ppbv	Liquid coolants/foams	10,800	10,200
HCFC-22	$CHClF_2$	11.9	Not naturally occurring	0.213 ppbv	Liquid coolants	5,280	1,760
Perfluoro methane PCF-14	CF_4	50,000	0*	0.079 ppbv	Production of aluminium	4,880	6,630
Sulphur hexa-fluoride	SF6	3,200	0*	0.007 ppbv	Dielectric fluid	17,500	23,500

ppmv = parts per million by volume
ppbv = parts per billion by volume
* trace amounts are found naturally

rise over the last century. Fourth, we have analysed the effects of natural changes on climate including sun spots and volcanic eruptions, and though these are essential to understanding the pattern of temperature changes over the last 150 years, they can not explain the overall warming trend (Figure 5). Fifth, we have observed significant changes in the Earth climate system including the retreat of sea ice in the Arctic, retreating mountain glaciers on all continents, and shrinking permafrost and increased depth of its active layer. Sixth, we continually track global weather and have seen significant shifts in the number and intensity of extreme events. In this chapter evidence for changes in global temperature, precipitation, sea level, and extreme weather events are examined.

Temperature

Temperatures for the last 2,000 years have been estimated from a number of sources, both direct thermometer-based and proxy-based indicators. What is a proxy? As used here and elsewhere, it is short for 'proxy variable'. The term 'proxy' is commonly used to describe a stand-in or substitute, as in 'proxy vote' or 'fighting by proxy'. In the same way, 'proxy variable' in the parlance of climatology means a measurable 'descriptor' that stands in for a desired (but unobservable) variable, such as past ocean or land temperature. So there is an assumption that you can use the proxy variable to estimate a climatic variable that you cannot measure directly. For example, infrared satellite measurements are examples of a proxy that can be used to estimating surface temperatures.

Thermometer-based measurements of air temperature have been recorded at a number of sites in North America and Europe as far back as 1760. The number of observation sites did not increase to sufficient worldwide geographical coverage to permit a global land average to be calculated until about the middle of the 19th century. Sea-surface temperatures (SSTs) and marine air

temperatures (MATs) were systematically recorded by ships from the mid-19th century, but even today the coverage of the Southern Hemisphere is extremely poor. All these data sets require various corrections to account for changing conditions and measurement techniques. For example, for land data each station has been examined to ensure that conditions have not varied through time as a result of changes in the measurement site, instruments used, instrument shelters, or the way monthly averages were computed, or the growth of cities around the sites, which leads to warmer temperatures caused by the urban heat island effect. In the IPCC science report, the influence of the urban heat island is acknowledged as real but negligible for the global temperature compilation (less than 0.006°C).

For SST and MAT, there are a number of corrections that have to be applied. First, up to 1941 most SST temperature measurements were made in seawater hoisted on deck in a bucket. Since 1941, most measurements have been made at the ships' engine water intakes. Second, between 1856 and 1910 there was a shift from wooden to canvas buckets, which changes the amount of cooling caused by evaporation that occurs as the water is being hoisted on deck. In addition, through this period there was a gradual shift from sailing ships to steamships, which altered the height of the ship decks and the speed of the ships, both of which can affect the evaporative cooling of the buckets. The other key correction that has to be made is for the global distribution of meteorological stations through time, which has varied greatly since 1870. But by making these corrections it is possible to produce a continuous record of global surface temperature from 1880 to 2012, which shows a global warming of 0.85°C over this period. These observations are supported by 50 years of balloon and satellite data. For example there are over 800 stations that twice a day release rawinsondes (meteorological instruments), or balloons, to measure temperature, relative humidity, and pressure through the atmosphere to a height of about 20 km, where they burst.

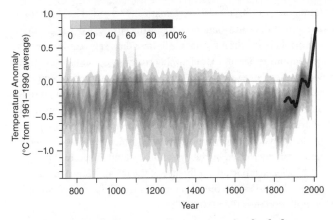

7. **Northern Hemisphere temperature reconstruction for the last 1,300 years**

Global temperatures have also been reconstructed for periods of time pre-dating instrumental or thermometer records. This is has been achieved by using palaeoclimate proxies such as the thickness of tree rings and the isotopic composition of ice cores or cave deposits to estimate local temperatures. The global warming 'hockey stick' term, which was discussed in Chapter 2, came about when the global mean temperature instrumental record was combined with palaeoclimate reconstructions of the last 2,000 years (see Figure 7). In this context it is clear that the last 50 years have been very different and much warmer than the previous 2,000 years.

Precipitation

There are two global precipitation data sets: Hulme and the Global Historical Climate Network (GHCN). Unfortunately, unlike for temperature, rainfall and snow data are not as well documented and the records have not been carried out for as long. It is also known that precipitation overland tends to be

underestimated by up to 10–15 per cent owing to the effects of airflow around the collecting dish. Without correction of this effect, there could be a spurious upward trend in global precipitation. Despite these problems there seems to be a significant increase of precipitation over the last 25 years (see Figure 8) particularly in the Northern Hemisphere middle latitudes. This is supported by evidence that since the 1980s atmospheric water content has increased over the land and ocean as well as in the upper troposphere. This is consistent with the extra water vapour that the warmer atmosphere can hold.

There is evidence for a global increase in precipitation but the evidence for this change is much stronger when considering individual regions. The latest IPCC report suggest that significant increases in precipitation have occurred in the eastern parts of North and South America, northern Europe, and northern and central Asia. It seems that seasonality of precipitation is also changing, for example in the high latitudes in the Northern Hemisphere, with increased rainfall in the winter and a decrease in the summer. Long-term drying trends have been observed on

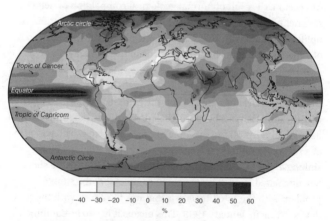

8. **Global precipitation changes (1900–2013)**

the Sahel, in the Mediterranean, southern Africa, and parts of southern Asia. It has also been observed that the amount of rain falling during heavy 'extreme' rain events has increased. These observations are supported by detailed compilation of all precipitation records for the Northern Hemisphere published in *Nature* in 2011 by Dr Seung-Ki Min and his colleagues, which showed a significant increase in the intensity of rainfall over the last 60 years.

Relative global sea level

The IPCC has also compiled all the current data on global sea level. It shows that between 1901 and 2010, the global sea level has risen by about 17 to 21 centimetres (cm) (see Figure 9). Sea-level change is difficult to measure, as relative sea-level changes have been derived from two very different data sets—tide-gauges and satellites. In the conventional tide-gauge system, the sea level is measured relative to a land-based tide-gauge benchmark. The major problem is that the land surface is much more dynamic that one would expect, with a lot of vertical movements, and that these become incorporated into the measurements. Vertical movements can occur as a result of normal geological compaction of delta sediments, the withdrawal of groundwater from coastal aquifers, uplift associated with colliding tectonic plates (the most extreme of which is mountain-building such as in the Himalayas), or ongoing postglacial rebound and compensation elsewhere associated with the end of the last ice age. The latter is caused by the rapid removal of weight when the giant ice sheets melted, so that the land that has been weighed down slowly rebounds back to its original position. An example of this is Scotland, which is rising at a rate of 3 millimetres (mm) per year, while England is still sinking at a rate of 2 mm per year, despite the Scottish ice sheet having melted 10,000 years ago. In comparison, the simple problem with the satellite data is that it is too short, with the best data starting in January 1993. This means it has to be combined with the tide-gauge data to look at long-term trends. However, the

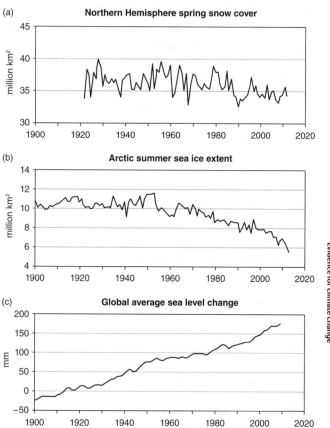

(a) **Northern Hemisphere spring snow cover**

(b) **Arctic summer sea ice extent**

(c) **Global average sea level change**

Year

9. Indicators of climate change

1993 to 2010 data clearly show more than a 60 mm rise in global sea level.

In summary, between 1901 and 2010 the global average sea level rose by 1.7 mm per year. While the fastest rise in sea level was observed between 1971 and 2010 of 3.2 mm per year. Between

1993 and 2010, the sea-level rise is made up of the following contributions: thermal expansion of the ocean contributed 1.1 mm per year (~39 per cent); Antarctic ice sheet 0.27 mm (~9 per cent); Greenland ice sheet 0.33 mm (~12 per cent); and glaciers and other ice caps 0.76 mm per year (~27 per cent); with approximately 0.38 mm per year (~13 per cent) from land water storage totalling 2.8 mm per year. These new data clearly show that the Greenland and Antarctic ice sheets have contributed to recent sea-level rise. At the moment it is estimated that Greenland is losing over 200 gigatonnes of ice per year, a six-fold increase since the early 1990s. While Antarctica is losing about 150 gigatonnes of ice per year, a five-fold increase since the early 1990s, and most of this loss is from the northern Antarctic Peninsula and the Amundsen sea sector of West Antarctica.

Other evidence for global warming

Other evidence for climate change comes from the high latitudes and from monitoring extreme weather events. The annual mean Arctic sea ice extent has decreased in total between 1979 to 2012 at a rate of 3.5 to 4.1 per cent per decade which means a loss of between 0.45 to 0.51 million km^2 per decade. Whereas the summer sea ice minimum has decreased even more by between 9.4 to 13.6 per cent per decade which is equivalent to 0.73 to 1.07 million km^2 per decade. In contrast between 1979 and 2012 the annual mean Antarctic sea ice extent has increased at a rate between 1.2 to 1.8 per cent per decade, which represents a growth of between 0.13 to 0.20 million km^2 per decade. There is also evidence from permafrost regions. Permafrost exists in high-latitude and high-altitude areas, where it is so cold that the ground is frozen solid to a great depth. During the summer months, only the top metre or so of the permafrost becomes warm enough to melt, and this is called the 'active layer'. There has been a 3°C warming in Alaska and 2°C warming in northern European/Russia down to at least a metre over the last 50 years, showing that the active layer has become deeper. The maximum area covered by seasonal

permafrost has decreased by 7 per cent in the Northern Hemisphere since 1900, with a decrease in the spring of up to 15 per cent. This increasingly dynamic cryosphere will amplify the natural hazards for people, structures, and communication links. Already we have seen this in damage to buildings, roads, and pipelines, such as the oil pipelines have been damaged in Alaska. In addition, there is evidence that most if not all non-ice sheet glaciers are in retreat. The amount of total snowfall and ice cover particularly in the Northern Hemisphere has great reduced (Figure 9). For example, ice cover records from the Tornio River in Finland, which have been compiled since 1693, show that the spring thaw of the frozen river now occurs a month earlier.

There is evidence too that our weather patterns are changing. For example, in recent years massive storms and subsequent floods have hit China, Italy, England, Korea, Bangladesh, Venezuela, and Mozambique. In England in 2000, 2007, and 2013/14, floods and storms classified as 'once-in-200-years events' have occurred within 13 years and frequently within a single year. Moreover, in Britain the winter of 2013/14 was the wettest six months since records began in the 18th century, while August 2008 was the wettest on record; and British spring is now coming earlier, with evidence of birds nesting 12 ± 4 days earlier than 35 years previously. Insect species—including bees and termites—that need warm weather to survive are moving northward, and some have already reached England by crossing the Channel from France. The frequency of heat waves has increased in Europe, Asia, and Australia—for example, in Europe in 2003 and 2007, Russia in 2010, USA in 2012, and Australia in 2009 and 2014. There is also evidence that more storms are occurring in the Northern Hemisphere. Wave height in the North Atlantic Ocean has been monitored since the early 1950s, from lightships, Ocean Weather Stations, and more recently from satellites. Between the 1950s and 1990s, the average wave height increased from 2.5 m to 3.5 m, an increase of 40 per cent. Storm intensity is the major determinant of wave height, which provides evidence for an increase in storm

activity over the last 40 years. This also fits with the observed increase in winter extra-tropical cyclones, that is those occurring in the mid-latitudes, which have increased markedly over the last 100 years, with significant rises in both the Pacific and Atlantic sectors since the early 1970s. There is also evidence for an increase in intense tropical hurricane activity since the 1970s in the North Atlantic.

What do the sceptics say?

One of the best ways to summarize the evidence for climate change is to review what the global warming sceptics or climate change deniers say against the current state-of-the-art science. Though I must stress I dislike the term sceptic because it seems to co-opt or steal a fundamental of science. All great scientists are sceptical about the state of current knowledge, and this drives them forward into making new observations and new experiments to allow them to develop new testable theories. The fundamental principle of 'weight of evidence' within science is a way of testing our sceptics about new ideas and data. So the term 'climate (change) deniers' is probably better term technically, as they are denying the scientific weight of evidence.

(1) Ice-core data suggest atmospheric carbon dioxide (CO_2) responds to global temperature, therefore atmospheric CO_2 cannot cause global temperature changes.

At the end of the last ice age as the Earth warmed up, we now know from ice cores from Greenland and Antarctica that the Northern and Southern Hemispheres warmed up at different times and at different rates. On top of this there are millennial-scale climate events, when huge amounts of ice broke off from the North American ice sheet flooding the North Atlantic Ocean with freshwater changing ocean circulation and in essence trying to push global climate back into much colder conditions.

One of these events called Heinrich event 1 occurred about 15,000 years ago and the other was the Younger Dryas, which occurred about 12,000 years ago. Because of the wonderfully named 'bipolar climate seesaw', whenever the Northern Hemisphere cools down heat is exported southwards and the Southern Hemisphere warms up. So if you compare an individual ice-core temperature record with reconstructed atmospheric CO_2 levels then there will be times when the relationship seems to swap. To really understand the relationship between global temperatures and CO_2, Dr Jeremy Shakun of Harvard University and colleagues created a master stack of all the temperature records across the end of the last ice age (see Figure 10). This shows that atmospheric CO_2 leads global temperatures adding to our confidence that it is contributing to the warming of the Earth as we exited the last great ice age.

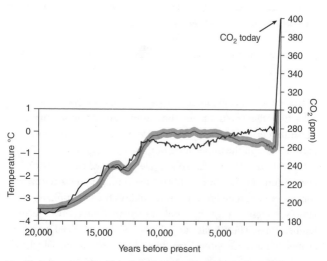

10. Global temperatures and carbon dioxide changes for the last 20,000 years

(2) Every data set showing global warming has been corrected or tweaked to achieve this desired result.

For people who are not regularly involved in science, this seems to be the biggest problem with the whole 'climate change has happened' argument. As shown above, all the climate data sets covering the last 150 years require some sort of adjustment. This, though, is part of the scientific process. For example, if great care had not been taken over the spurious trends in the global precipitation database we would now assume that global precipitation was increasing. Moreover, as science moves forward incrementally, it gains more and more understanding and insight into the data sets it is constructing. This constant questioning of all data and interpretations is the core strength of science: each new correction or adjustment is due to a greater understanding of the data and the climate system, and thus each new study adds to the confidence that we have in the results. This is why the IPCC report refers to the 'weight of the evidence', since our confidence in science increases if similar results are obtained from very different sources.

(3) Recent changes in global temperatures are due to changes in the Sun.

Both sceptics and climate scientists agree that sunspots and volcanic activity do influence climate and global temperatures. The difference between the two camps is that the sceptics put more weight on the importance of these natural variations. There is evidence that the 11-year solar cycle, during which the Sun's energy output varies by roughly 0.1 per cent, can influence ozone concentrations, temperatures, and winds in the stratosphere. However, these changes have only a very small effect on surface temperatures. There is also no evidence of a pronounced long-term change in the Sun's output over the past century, during which time human-induced increases in CO_2 concentrations have been significant. Figure 11 shows that since 1980, as global temperatures have increased, there has been no discernable

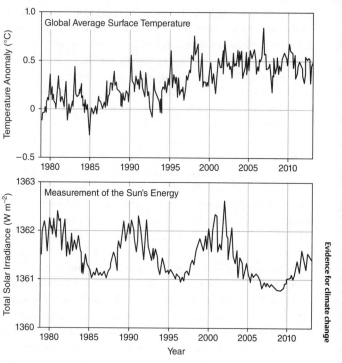

11. Sunspot and global temperatures

change in trend in the solar output. Moreover, the observation of different temperature trends at different altitudes in the atmosphere clearly show that overall warming cannot be due to increased solar radiation.

(4) The recent slow-down in warming means that climate change is not as bad as predicted.

One of the reasons that it is perceived that global warming has slowed down or even stopped is because of the prominence of 1998, which was one of the warmest years on record. This year was

extreme because it followed the strong 1997–8 El Niño event which adds at least 1°C warming around the world. However, when one observes the global temperature graph, one's eye is drawn to this event and it looks like there has not been any warming since then. It is clear that the increase in average surface temperature has slowed compared to the previous decade. So the rate of warming was slower in the 2000s compared with the 1990s but the 2000s are still significantly warmer than the 1990s (Figure 5). A short-term slowdown in the warming of the Earth's surface such as this does not invalidate our understanding of climate change. Changes in the rate or warming between decades occur naturally in the climate system. This can be seen in the observations of temperatures over the past 150 years. This is because the atmosphere stores very little heat, and so surface temperatures can be rapidly affected by heat uptake elsewhere in the climate system and by changes in external influences on climate such as volcanic eruptions and sunspots. More than 90 per cent of the heat added to Earth is absorbed by the oceans and penetrates only slowly into deep water. A faster rate of heat penetration into the deeper ocean will slow the warming seen at the surface and in the atmosphere, but by itself will not change the long-term warming that will occur due to particular levels of GHGs. For example, we know that heat comes out of the ocean into the atmosphere during warm El Niño events, and more heat is pumped into the ocean depths during cold La Niñas. Over the last decade there has been coincidence of a number of small cooling effects including a relatively quiet period of solar activity (Figure 11) and a measured increase in the amount of aerosols (reflective particles) in the atmosphere due to the cumulative effects of a succession of small volcanic eruptions. But when the global temperature record is looked at decade by decade it is clear that the last three decades were all successively warmer than the previous one.

Summary

Over the last 150 years, significant changes in climate have been recorded, which are markedly different from the last at least 2,000

years. These changes include a 0.85°C increase in average global temperatures, sea-level rise of over 20 cm, significant shifts in the seasonality and intensities of precipitation, changing weather patterns, and the significant retreat of Arctic sea ice and nearly all continental glaciers. According to the US National Oceanic and Atmospheric Administration between 1880 and the beginning of 2014, the 13 warmest years on record have all occurred within the last 16 years, with 2010 the warmest year, followed by 2005, 1998, 2003, 2013, 2002, 2006, 2007, 2009, 2004, 2012, 2001, and 2011. The IPCC 2013 report states that the evidence for climate change is unequivocal and there is very high confidence that this warming is due to human emissions of GHGs. This statement is supported by six main lines of evidence: (1) the rise in GHGs in the atmosphere has been measured and the isotopic composition of the gases shows that the majority of the additional carbon comes from the burning of fossil fuels; (2) laboratory and atmospheric measurements show that these gases absorb heat; (3) significant changes in global temperatures and sea-level rise have been observed over the last century; (4) other significant changes have been observed in cryosphere and atmosphere including retreating sea ice and glaciers, and extreme weather events; (5) there is clear evidence that natural processes including sun spots and volcanic eruptions can not explain the warming trend over the last 100 years; and (6) we now understand the longer term past climate changes and the role GHGs have played in regulating the climate of our planet.

Chapter 4
Modelling future climate

The whole of human society operates on knowing the future weather. For example, farmers in India know when the monsoon rains will come next year and so they know when to plant the crops. Farmers in Indonesia know there are two monsoon rains each year, so next year they can have two harvests. This is based on their knowledge of the past, as the monsoons have always come at about the same time each year in living memory. But the need to predict goes deeper than this as it influences every part of our lives. Our houses, roads, railways, airports, offices, cars, trains, and so on are all designed for the local climate. For example, in England all the houses have central heating, as the outside temperature is usually below 20°C, but no air-conditioning as temperatures rarely exceed 26°C. While in Australia the opposite is true and most houses have air-conditioning but rarely central heating. Predicting future climate is, therefore, essential, as we know that we can no longer rely on records of past weather of an area to tell us what the future will hold. We have to develop new ways of predicting the future, so that we can plan our lives and so that human society can continue to function fully. So we have to model the future.

Models

There is a whole hierarchy of climate models, from relatively simple box models to extremely complex three-dimensional GCMs. Each has a role in examining and furthering our understanding of the global climate system. However, it is the complex three-dimensional general circulation models that are used to predict future global climate. These comprehensive climate models are based on physical laws represented by mathematical equations, which are solved using a three-dimensional grid over the globe. To obtain the most realistic simulations, all the major parts of the climate system must be represented in sub-models, including atmosphere, ocean, land surface (topography), cryosphere, and biosphere, as well as the processes that go on within them and between them. Most global climate models have at least some representation of each of these components. Models that couple together both the ocean and atmosphere components are called atmosphere–ocean general circulation models (AOGCMs).

Over the last 30 years there has been a huge improvement in climate models. This has been due to our increased knowledge of the climate system but also because of the nearly exponential growth in computer power. There has been a massive improvement in spatial resolution of the models from the very first IPCC report in 1990 to the latest in 2013. The current generation of AOGCMs have multiple layers in the atmosphere, land, and ocean and a spatial resolution greater than one point every 100 km by 100 km (see Figure 12). Equations are typically solved for every simulated 'half-hour' of a model run. Many physical processes, such as cloud and ocean convection, of course take place on a much smaller scale than the model can resolve. Therefore, the effects of small-scale processes have to be lumped together, which is referred to as 'parameterization'. Many of these parameterizations are, however, checked with separate 'small-scale-process models' to validate the scaling up of these smaller

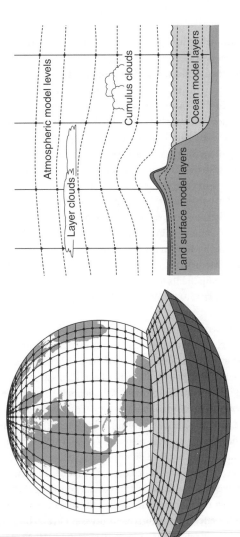

12. Generic structure of a global climate model

influences. The reason that the spatial scale is limited is that comprehensive AOGCMs are very complex and use a huge amount of computer time to run. At the moment, much of the improvement in computer processing power that has occurred over the last decade has been used to improve the representation of the global climate system by coupling more models directly into the GCMs. The very latest models or 'climate simulators', as some groups are now referring to them, include much better representations of atmospheric chemistry, clouds, aerosol processes, and the carbon cycle, including land vegetation feedbacks. But the biggest unknown in the models is not the physics, it is the estimation of future global GHG emissions over the next 90 years. This includes many variables, such as the global economy, global and regional population growth, development of technology, energy use and intensity, political agreements, and personal lifestyles. Hence you could produce the most complete model in the world, taking two years to simulate the next 100 years, but you would have only one prediction of the future, based on only one estimate of future emissions—which might be completely wrong. Individual models are therefore run many times with different inputs to provide a range of changes in the future. In fact, the latest IPCC science report has consulted the results of multiple runs from over 40 different AOGCMs to provide the basis for their predictions. Of course, as computer processing power continues to increase, both the representation of coupled climate systems and the spatial scale will continue to improve.

Not only does the IPCC fifth assessment include some significant improvements in the presentation of the physical processes of the climate system but also many of the models had a small increase in spatial resolution. The models also focus on decadal forecasts to help understand the internal variability of the climate to understand when the rate of warming may slow down or speed up. There are also separate chapters dealing with near-term climate up to 2050 and climate change after 2100.

Carbon cycle

At the heart of the AOGCMs is the carbon cycle and estimating what happens to anthropogenic carbon dioxide and methane emissions. As about half of all our carbon emissions are absorbed by the natural carbon cycle and do not end up in the atmosphere, but rather in the oceans and the terrestrial biosphere, the Earth's carbon cycle is complicated, with both sources and sinks of carbon dioxide. Figure 13 shows the global carbon reservoirs in gigatonnes or 1,000 million tonnes (GtC) and fluxes (the ins and outs of carbon in GtC per year). These indicated figures show the changes since the industrial revolution. Evidence is accumulating that many of the fluxes can vary significantly from year to year.

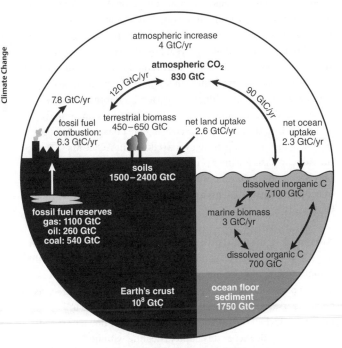

13. **The carbon cycle, in gigatonnes of carbon (GtC)**

This is because in contrast to the static view conveyed in figures like this one, the carbon system is dynamic, and coupled to the climate system on seasonal, inter-annual, and decadal timescales. What has become clear is that the surface ocean and the land biosphere both take up about 25 per cent each of our carbon emission every year. However, as the oceans continue to warm they can hold less dissolved carbon dioxide, which means that their uptake will reduce. Also as we continue to deforest and substantially alter land use then the land biosphere ability to absorb carbon diminishes.

Warming and cooling effects

As well as the warming effects of the GHGs, the Earth's climate system is complicated in that that there are also cooling effects (see Figure 14). This includes the amount of particles in the air (which are called aerosols, many of which come from human pollution such as sulphur emissions from power stations) and these have a direct effect on the amount of solar radiation that hits

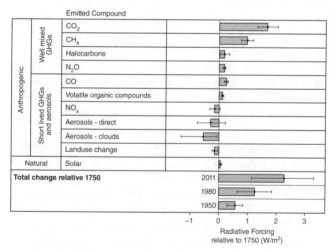

		Emitted Compound	
Anthropogenic	Well mixed GHGs	CO_2	
		CH_4	
		Halocarbons	
		N_2O	
	Short lived GHGs and aerosols	CO	
		Volatile organic compounds	
		NO_x	
		Aerosols – direct	
		Aerosols – clouds	
		Landuse change	
Natural		Solar	
Total change relative 1750		2011	
		1980	
		1950	

Radiative Forcing
relative to 1750 (W/m²)

14. **Radiative forcings between 1750 and 2011**

the Earth's surface. Aerosols may have significant local or regional impact on temperature. In fact, the AOGCMs have now factored them into the computer simulations of climate change, and they provide an explanation of why industrial areas of the planet have not warmed as much as previously predicted. Water vapour is a GHG, but, at the same time, the upper white surface of clouds reflects solar radiation back into space. This reflection is called 'albedo'—and clouds and ice have a high albedo, which means that they reflect large quantities of solar radiation from surfaces on Earth. Increasing aerosols in the atmosphere also increases the amount of clouds as they provide points on which the water vapour can nucleate. Predicting what will happen to the amount and types of clouds, and their warming potential, has been one of the key challenges for climate scientists.

Emission models of the future

A critical problem with trying to predict future climate is predicting the amount of carbon dioxide emissions that will be produced in the future. This will be influenced by population growth, economic growth, development, fossil-fuel usage, the rate at which we switch to alternative energy, the rate of deforestation, and the effectiveness of international agreements to cut emissions. Out of all the systems that we are trying to model into the future, humanity is by far the most complicated and unpredictable. If you want to understand the problem of predicting what will happen in the next 100 years, imagine yourself at the beginning of the 20th century and what you would have predicted the world to be like in the 21st century. At the beginning of the 20th century, the British Empire was the dominant world power due to the industrial revolution and the use of coal. Would you have predicted the switch to a global economy based on oil after the Second World War? Or the explosion of car use? Or the general availability of air travel? Even 20 years ago, it would have been difficult to predict that there would be budget airlines, allowing for cheap flights throughout Europe and the USA.

The original IPCC reports used simplistic assumption of GHG emissions over the next 100 years. From 2000 onwards the climate models used the more detailed Special Report on Emission Scenarios. The 2013 IPCC Fifth Assessment report used more sophisticated Representative Concentration Pathways (RCPs) which considered a much wider variable input to the social-economic models including population, land use, energy intensity, energy use, and regional differentiated development (see Table 2). However, the new RCPs mean that detailed comparison of the 2013 IPCC results is difficult with the IPCC 2001 and 2007 outputs, which used the SRES.

There are four main RCPs that are used, defined by the final radiative forcing achieved by the year 2100, and they range from 2.6 to 8.5 watts per square metre (W/m^2). Radiative forcing is defined as the difference of sunlight (radiant energy) received by

Table 2. Defining Representative Concentration Pathway used in the IPCC 2013 report

Representative Concentration Pathway (RCP)	Description
RCP8.5	Rising radiative forcing pathway leading to 8.5 W/m^2 (~1370 ppm CO_2 eq*) by 2100.
RCP6	Stabilization without overshoot pathway to 6 W/m^2 (~850 ppm CO_2 eq) at stabilization after 2100
RCP4.5	Stabilization without overshoot pathway to 4.5 W/m^2 (~650 ppm CO_2 eq) at stabilization after 2100
RCP2.6 (also called RCP3PD)	Peak in radiative forcing at ~3 W/m^2 (~490 ppm CO_2 eq) before 2100 and then decline to 2.6 W/m^2 by 2100 (~420 ppm CO_2 eq).

* CO_2 eq is the carbon dioxide equivalent of all the GHGs radiative forcings combined

the Earth and the energy radiated back to space. Radiative forcing is quantified at the tropopause, which is the lowest layers of the Earth's atmosphere where all weather occurs. Its height ranges from 10 km (~6 miles) at the Poles to nearly 18 km (~11 miles) in the Tropics. Radiative forcing is measured in units of W/m^2 of the Earth's surface. A positive forcing (more incoming energy) warms the system, while negative forcing (more outgoing energy) cools it. The radiative forcing of Earth can change due to changes in insolation (incident solar radiation) and the concentrations of GHGs and aerosols. The four RCPs were selected to be representative of the three most likely emission pathways: two medium stabilization scenarios (RCP4.5/RCP6); and one business as usual baseline emission scenarios (RCP8.5). A RCP was also included to illustrate what could be achieved if every mitigation strategy was employed (RCP2.6). This pathway is also referred to as RCP3PD, a name that emphasizes the radiative forcing trajectory as it first goes to a peak forcing level of 3 W/m^2 followed by a decline, the PD representing peak then decline (see Figure 15). This shows that emissions would initially increase producing a radiative forcing of 3 W/m^2 and then there would be huge cut backs in emissions so that by the year 2100 a radiative forcing of only 2.6 W/m^2 was achieved. What is rarely mentioned about this RCP is that it assumes negative emissions from 2070 onwards, which means that not only does the world have to cease producing any carbon emissions by 2070 but that after this date we will actively be taking carbon dioxide and other GHGs out of the atmosphere, which is an immense undertaking.

Modelling uncertainty

In the most recent IPCC Fifth Assessment report, the RCPs were inputted into about 40 GCMs. Each of these models has their own independent design and parameterizations of key processes. The independence of each model is important, as confidence may be derived from multiple runs on different

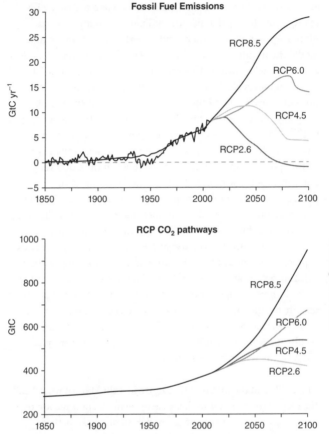

15. Future carbon emission scenarios

models providing similar future climate predictions. In addition
the differences between the models can help us to learn about
their individual limitations and advantages. Within the IPCC,
due to political expediency, each model and its output is
assumed to be equally valid. This is despite the fact that some

are known to perform better than others when tested against reality provided by the historic and palaeoclimate records. Moreover though we understand uncertainty within a single model the notion of quantifying uncertainty from many models currently lacks any real theoretical background or basis. The IPCC combines all the models used for each run and then presents the mean and the uncertainty between the models. This way it is clear that there are difference in the model output but that in general they agree and show very different futures based on which RCP we take. The uncertainties in the IPCC 2013 report are slightly higher than those in the 2007 report and this is because of our greater understanding of the processes and our ability to quantify that knowledge. So though our confidence in the climate models has increased, so has the range of possible answers for any specific GHG forcing. Dr Dan Rowlands (Oxford University) and colleagues in 2012 recently explored the amount of uncertainty inherent in complex models by running one specific climate model through nearly 10,000 simulations (as opposed to the usual handful of runs that can be usually be managed). While his average results matched well with the IPCC projections, they found that more extreme results, including warming of up to 4°C by 2050, were just as likely as the less extreme results.

The RCPs and the different GCMs are just the start of what I call the cascade of uncertainty, because these model outputs are then used in much higher resolution models to provide a better understanding of the potential impacts of climate change (see Figure 16). This is called down-scaling and is a huge problem recognized in the modelling community, because precipitation is spatially and temporally highly variable but essential to model if human impacts are to be understood. Ultimately the cascade of uncertainty leads to a huge range of potential futures at a regional level that are in some cases contradictory. For example, detailed hydrological modelling of the Mekong River Basin using climate model input from just a single GCM by the UK Met Office

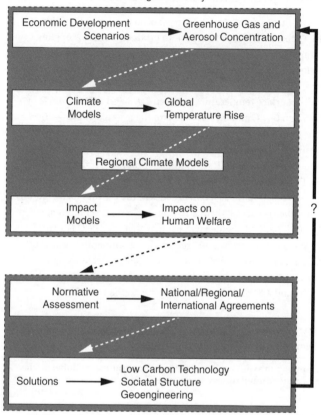

Economic Development Scenarios → Greenhouse Gas and Aerosol Concentration

Climate Models → Global Temperature Rise

Regional Climate Models

Impact Models → Impacts on Human Welfare

?

Normative Assessment → National/Regional/International Agreements

Solutions → Low Carbon Technology Sociatal Structure Geoengineering

Modelling future climate

16. **Cascading uncertainty through climate change models and policy**

(HadCM3) lead to projected future changes in annual river discharge ranging from a decrease of 5.4 per cent to an increase of 4.5 per cent. Changes in predicted monthly discharge are even more dramatic ranging from –16 per cent to +55 per cent. Advising policy makers becomes extremely hard when the uncertainties do not even allow one to tell if the river catchment system in the future will have more or less water.

Future global temperatures and sea level

Between 32 and 39 AOGCMs have been run for each of the RCPs for the IPCC 2013 report, to produce scenarios of global temperature and sea level changes that may occur by 2100. This is a significant change from the IPCC 2001 report, in which only seven models were used. These climate models suggest that the global surface temperature between 2016 and 2035 will rise by between 0.3°C and 0.7°C relative to the average of 1986–2005. The global temperatures rise for the average of 2081–2100 again relative to the average of 1986–2005 will be heavily dependent on the RCP we follow (see Table 3). If the realistic RCPs are considered the global temperatures could rise between 1.1 and 4.8°C in the last two decades of the century (see Figure 17). With the rise of 0.8°C already, this would represent a total rise of 1.9°C to 5.6°C. An added confusion is that the IPCC Fourth Assessment report in 2007 report global temperatures at 2100 instead of an average of 2081–2100. They reported using the best estimates for the original six emission scenarios, a range between 1.8°C and 4°C by 2100. This compares to the IPCC Fifth

Table 3. Temperature and sea level projection by Representative Concentration Pathway

Representative Concentration Pathway	Global temperature change (°C) 2081–2100	Global sea level rise (m) 2081–2100
RCP8.5	2.6 to 4.8 (mean 3.7)	0.45 to 0.82 (mean 0.63)
RCP6	1.4 to 3.1 (mean 2.2)	0.33 to 0.63 (mean 0.48)
RCP4.5	1.1 to 2.6 (mean 1.8)	0.32 to 0.63 (mean 0.47)
RCP2.6 or RCP3PD	0.3 to 1.7 (mean 1.0)	0.26 to 0.55 (mean 0.40)

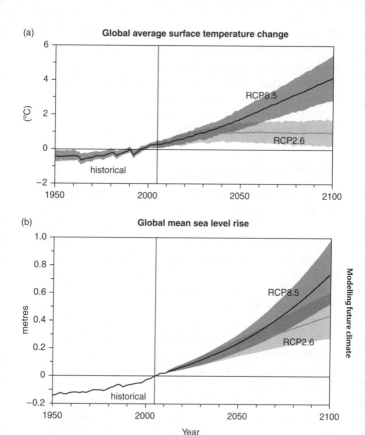

17. Global temperatures, Arctic sea ice, and sea level in the 21st century

Assessment Report 2013's final mean global temperature for
RCP8.5 and RCP4.5 of between 1.9°C and 4.1°C by 2100. A
strikingly similar set of results.

In terms of sea-level rise, again this is dependent on which RCP
we follow, but taking the three realistic ones then it will be
between 0.32 m and 0.82 m in the last two decades of the century

(Table 3 and Figure 17). With the rise of 20 cm, which has already occurred then this would represent a total rise of 0.52 m and 1.02 m. If we look at the final projected sea level at 2100, they show an increase in global mean sea level of between 27 cm and 98 cm. This is similar but more extreme that the projection made by the IPCC 2007 report, which suggests a sea level rise of between 28 cm and 79 cm by 2100.

Modelling extreme events

Climate change modelling has advanced so rapidly in the last decade that it can now attempt to attribute the contribution of anthropogenic climate change to extreme weather events. A few years ago this would be unheard of, and the standard communication line was that scientist could not attribute individual weather events to climate change, but the event in question may be consistent with what is expected to happen in the future. However, with increased computer power it is possible to run regional climate scenarios thousands of times with and without the contribution of anthropogenic GHGs and so assess the potential impact on the occurrence of extreme weather events. A discernable contribution of anthropogenic climate change has been found for UK floods in 2000, the Russian heat wave of 2010, and the Texan and East African droughts of 2011; while no climate change influence has been found for the floods in Thailand in 2011 or in Pakistan in 2010. This science though is still in its infancy and throws up sometimes contradictory studies due to being as yet unable to define what we mean exactly by an anthropogenic climate change contribution. This complexity is shown by two research papers published relating to the Russian heat wave of 2010; as: one concluded that climate change had not contributed to the event while the other concluded that it had. This apparent mismatch was caused by the papers asking different questions. The first study showed that climate change had had little or no effect on the magnitude of the Russian heat wave, while the second study showed that climate change had increased the

frequency at which these events could occur three-fold. This demonstrates the importance of scientists and policy makers asking the right questions.

What the sceptics say

One of the best ways to summarize the perceived problems of modelling climate change is to review what the sceptics say.

(1) Clouds can have negative feedbacks on global climate which will reduce the effects of climate change

As has been the case since the very first IPCC report in 1990, the greatest uncertainty in future predictions is the role of the clouds and their interaction with radiation. Clouds can both absorb and reflect radiation, thereby cooling the surface, and absorb and emit long-wave radiation, thus warming the surface. The competition between these effects depends on a number of factors: height, thickness, and radiative properties of clouds. The radiative properties and formation and development of clouds depend on the distribution of atmospheric water vapour, water drops, ice particles, atmospheric aerosols, and cloud thickness. The physical basis of how clouds are represented or parametrized in the AOGCMs has greatly improved through the inclusion of representations of cloud microphysical properties in the cloud water budget equations. Clouds still represent a significant source of uncertainty in climate simulations. However, as Figure 14 shows, even if the most extreme cooling value is applied for clouds, the warming factors are still three times larger.

(2) Different models give different results, so how can we trust any of them?

This is a frequent response from many people not familiar with modelling, as there is a feeling that somehow science must be able to predict an exact future. However, in no other walk of life do we

expect this precision. For example, you would never expect to get a perfect prediction of which horse will win a race or which football team will emerge triumphant. The truth is that none of the climate models is exactly right. But what they provide is the best estimate that we have of the future. Now this view of the future is strengthened by the use of more than one model, because each model has been developed by different groups of scientists around the world, using different assumptions and different computers, and thus they provide their own particular future prediction. What causes scientists to have confidence in the model results is that they all roughly predict the same trend in global temperature and sea level for the next 100 years. One of the great strengths of the 2013 IPCC report is that it used over 40 models, while the 2007 IPCC report used 23 international models, compared to seven in 2001. Another strength of this large-scale multiple model approach is that scientists can also give an estimation of how confident they are in the model results and also a range of possible predictions, discussed earlier. One key test of climate models is the equilibrium climate sensitivity (ECS) whereby the model predicts what the global temperature change would be if pre-industrial carbon dioxide levels were doubled. These results have been very consistent over the last 40 years (see Figure 18) and the 2013 IPCC report suggests the range is between 1.5°C and 4.5°C, which is consistent with other measures. What is even more amazing is that in 2014 the UK climate sceptical think tank the Global Warming Policy Foundation (GWPF) published its own report, which looked into ECS. They concluded that ECS was between 1.25°C and 3.0°C with a best estimate of 1.75°C. Though this is lower than the IPCC estimate it is still a major breakthrough that an organization such as GWPF is now recognizing the impacts of carbon emissions on the atmosphere.

(3) Climate models fail to predict weather.

Many people get weather and climate confused. Climate is generally defined as the average weather. The original definition of

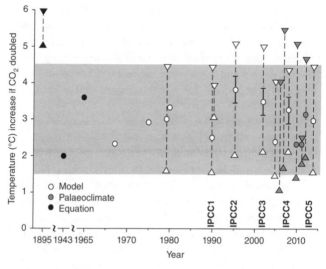

18. Equilibrium climate sensitivity

climate was the average weather over 30 years, this has been
changed because we now know that our climate is changing and
significant changes have been seen every decade for the last 50
years. It is the chaotic nature of weather that can make it
unpredictable beyond a few days, as the Earth's climate system is
sensitive to extremely small perturbations in initial conditions.
For example, extremely slight changes in air pressure over the
USA have an influence on the direction and duration of a
hurricane. Though there have been amazing advances in weather
prediction over the last decade, leading to storm warnings seven
days in advance instead of two in the 1980s. Climate modelling is,
however, much easier as you are dealing with long-term averages.
A good comparison is that though it is impossible to predict at
what age any particular person will die, we can say with a high
degree of confidence that the average life expectancy of a person
in a developed country is about 80 years. The modelling of climate
is not limited in the same way as the prediction of the weather

because the longer term systematic influences on the atmosphere are not reliant on the initial conditions. So the longer term trends in regional and global climate are not controlled by small-scale influences. Also, as described above we now have the computer power to go back and test whether extreme weather events were more intense or frequent due to anthropogenic climate change.

(4) Climate models fail to reconstruct or predict natural variability.

The global climate system contains cyclic variations, which occur on a decade or sub-decade timescale. The most famous is El Niño, which is a change in both ocean and atmospheric circulation in the Pacific region occurring every three to seven years that has a major influence on the rest of the global climate. Sceptics argue that climate models have been unable to simulate satisfactorily these events in the past. However, climate models have become better at reconstructing these past variations in El Niño–Southern Oscillation (ENSO), North Atlantic Oscillation (NAO), and related Arctic Oscillation (AO) as there has been an increasing realization that these have a profound impact upon regional climate (see *Climate: A Very Short Introduction*). Most models are able to depict these natural variations, picking out particularly the 1976 climate shift that occurred in the Pacific Ocean. All the AOGCMs have predicted outcomes for ENSO and NAO for the next 100 years. However, a lot of improvement is required before there will be confidence in the model predictions. It is, though, testament to the realism of the AOGCMs that they can indeed reconstruct and predict future trends in these short-term oscillations.

(5) Climate models cannot reconstruct past climate.

Past climates are an important test for global climate models and the IPCC Fifth Assessment report has a whole chapter dedicated to palaeoclimatology. The biggest climate shift, for which we have many palaeoclimate reconstructions, is that of the last ice age, which ended about 10,000 years ago. A comparison between

palaeoclimate data for the most extreme stage of the ice age, which occurred 18,000 years ago, suggests that the global climate models are rather good. It shows that the AOGCMs used for predicting future climate can do a good job of reconstructing the extreme conditions of an ice age and can get sea level close to 120 m lower, and global temperatures 6°C cooler, with atmospheric carbon dioxide one-third lower and atmospheric methane halved. One important observation is that the models are conservative, and they systematically underestimated the extremes of the last ice age. This means we can assume that the future climate predictions are also conservative, and thus climate change is very likely to be at the top end of the estimates.

(6) What about Galactic cosmic rays (GCRs).

GCRs are high-energy particles that cause ionization in the atmosphere and it has been suggested that they could affect cloud formation. GCRs vary inversely with solar variability because of the effect of solar wind. This is an excellent example of how climate science progresses by gaining new knowledge, testing it thoroughly, and if required adding it into the climate models. However, there seems to be no correspondence with the variations of cosmic rays and the global total cloud cover since 1991 or to global low-level cloud cover since 1994. Together with the lack of a proven physical mechanism and the plausibility of other causal factors affecting changes in cloud cover, this makes the association between GCR-induced changes in aerosol and cloud formation unlikely. Some colleagues have found that the evidence showed that connections between solar variation and climate were more likely to be mediated by direct variation of insolation rather than cosmic rays, and concluded that varying solar activity, either by direct solar irradiance or by varying cosmic ray rates, would have contributed less than 0.07°C warming since 1956, in other words less than 14 per cent of the observed global warming. Therefore, a review of the recent and historical literature continues to find that the link between cosmic rays and climate is tenuous.

Summary

Modelling future climate change is about understanding the fundamental physical processes of the climate system. Four new emission scenarios were produced for the 2013 IPCC Science report using a much wider set of input to the socioeconomic models including population, land use, energy intensity, energy use, and regional differentiated development. One of these emissions pathways (RCP2.6) was developed to indicate to policy makers what could be achieved in terms of climate change if all possible mitigation strategies were employed as soon as possible. Over 40 climate models were used in developing the IPCC projections and the quantification of uncertainty, providing a huge 'weight of evidence'. Using the three main realistic carbon emissions pathways over the next 85 years, the climate models suggest the global mean surface temperature could rise by between 2.8°C and 5.4°C by 2100. However it must remembered that global temperatures will stop changing once we get to year 2100. Figure 19 shows

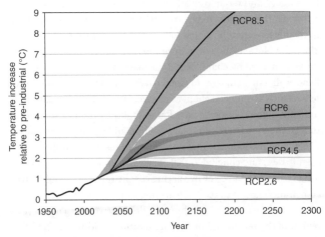

19. **Global surface temperatures (1950–2300)**

how temperatures could continue to rise way beyond the levels of this century depending on the chosen emission pathways. Using the three main realistic carbon emissions pathways the models also predict an increase in global mean sea level of between 52 cm and 98 cm by 2100.

Chapter 5
Climate change impacts

This chapter assesses the potential impacts of climate change and how these change scale and intensify with increasing warming. The IPCC 2014 'Impacts, Adaptation and Vulnerability' report looks at potential impacts on a continental level as well as by different sectors, such as freshwater resources, ecosystems, coastal and ocean systems, food security, and human health. It is also necessary to estimate the extent and magnitude of climate change at national and local levels. There are a number of excellent national reports and tools, such as the US National Climate Assessment and the UK Climate Impacts Programme, both of which have interactive tools to understand the potential effects of climate change within their own countries. In this chapter the potential impacts are broken down in to sectors: coast, storms and floods, heat waves and droughts, human health, biodiversity, ocean acidification, and agriculture.

What is dangerous climate change?

One of the most important questions for policy makers is what is dangerous climate change? Of course, this does depend on where you live. For example, if you are in one of the small island nations, any sea-level rise could be considered dangerous because it directly

results in loss of land. However, looking at the bigger picture, if we are to cut global GHG emissions we need a realistic target concerning the degree of climate change with which we can cope. In February 2005, the British government convened an international science meeting at Exeter, UK, to discuss this very topic. This was a very political science meeting, as the UK government was looking for a recommendation to take to the G8 meeting in Gleneagles. At that time Britain held both the chair of the G8 and presidency of the EU, and the then prime minister Tony Blair wished to push forward internationally his joint agenda of climate change reduction and poverty alleviation in Africa. The meeting did come up with a 'magic number' of 2°C above pre-industrial average temperature: below this threshold, there seem to be both winners and losers due to regional climate change, but above this figure everyone seems to lose. It wasn't just this particular meeting that has come up with the magic 2°C limit; many other researchers, including at the IPCC, have arrived at similar conclusions from very different backgrounds and starting assumptions. But it should always be remembered that this is a *political* number, as the definition of what is dangerous climate change is a societal rather than scientific decision. Figure 20 shows the numbers of people that potentially could be at risk from water shortages, hunger, malaria, and flooding by 2080. In Figure 20 it seems 2°C is where the numbers increase radically, particularly for water shortages, so 2°C has become a powerful and important symbol of the challenges facing human society. The major problem is that it is unlikely we can keep global temperature increases down to 2°C, as we have seen that temperature have already increase by 0.8°C, and because of the inertia in the climate system by 2035 we already expect a rise of between 0.3°C and 0.7°C. So without doing anything, we will be up to 1.1–1.5°C by 2035.

Extreme events and society's coping range

The single biggest problem with climate change is our inability to predict the future. Humanity can live, survive, and even flourish in

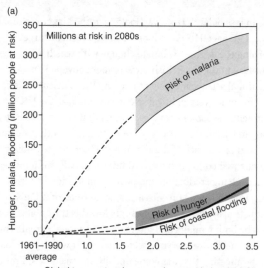

(a)

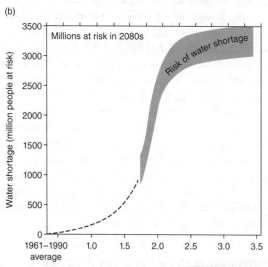

(b)

20. Climate change risks as a function of increasing global temperatures

extreme climates from the Arctic to the Sahara, but problems arise when the predictable extremes of local climate are exceeded. For example, heat waves, storms, droughts, and floods in one region may be considered fairly normal weather in another. This is because each society has a coping range, a range of weather with which it can deal. Figure 21 shows the theoretical effect of combining the societal coping range with climate change. In our present climate, the coping range encompasses nearly all the variation in weather with maybe only one or two extreme events. These could be one-in-200-year events that surpass the ability of that society to copy with them. As the climate moves gently to its new average, if the coping range stays the same then many more extreme events will occur. Hence a one-in-200-year event may become a one-in-50-year event. The good news is that the societal coping range is flexible and can change to cope with the shifting baseline and the more frequent extreme events; as long as there is strong climate science to provide clear guidance on what sort of changes are going to occur. The speed with which the societal coping range can vary depends on what aspect of society is being affected. So adaptation of the individual's behaviour can be extremely quick, while building major infrastructure can take decades to complete. One of the biggest challenges of climate change is to build flexible and resilient societies that are able to adapt to a changing future.

Coasts

As we have seen, the IPCC reports that sea level could rise by between 27 cm and 98 cm by 2100. This prediction is of major concern to all coastal areas, as rising sea levels will reduce the effectiveness of coastal defences against storms and floods, and increase the instability of cliffs and beaches. In Britain, the USA, and the rest of the developed world, the response to this danger has been to add another few feet to the height of sea walls around property on the coast, the abandoning of some poorer quality agricultural land to the sea (as it is no longer worth the expense

Climate Change

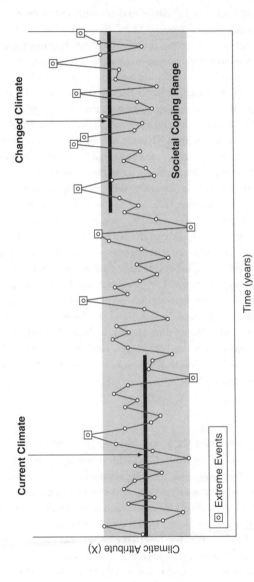

21. Climate change, societal coping range, and extreme events

of protecting it), and the enhancement of legal protection for coastal wetlands, being nature's best defence against the sea. However, globally, there are some nations based on small islands and river deltas that face a much more urgent situation (see Figure 22).

For small island nations, such as the Maldives in the Indian Ocean and the Marshall Islands in the Pacific, a 1 m rise in sea level would flood up to 75 per cent of the dry land, making the islands uninhabitable. Interestingly, it is also these countries, which rely mainly on tourism, that have some of the highest fossil-fuel emissions per head of population. Other major concentrations of population who are at risk are those live by river deltas, including Bangladesh, Egypt, Nigeria, and Thailand. A World Bank report concluded that human activities on the deltas, such as dams and fresh-water extraction, were causing these areas to sink much faster than any predicted rise in sea level, increasing their vulnerability to storms and floods.

In the case of Bangladesh, over three-quarters of the country is within the deltaic region formed by the confluence of the Ganges, Brahmaputra, and Meghna rivers. Over half the country lies less than 5 m above sea level; thus flooding is a common occurrence. During the summer monsoon a quarter of the country is flooded. Yet these floods, like those of the Nile, bring with them life as well as destruction. The water irrigates and the silt fertilizes the land. The fertile Bengal delta supports one of the world's most dense populations, over 110 million people in 140,000 square kilometres (km^2). Every year, the Bengal delta should receive over one billion tonnes of sediment and a 1,000 cubic kilometres (km^3) of fresh water. This sediment load balances the erosion of the delta both by natural processes and human activity. However, the Ganges, Brahmaputra, and Meghna have been dammed for irrigation and power generation preventing the movement of silt down river. The reduced sediment input is causing the delta to subside. Exacerbating this is the rapid extraction of fresh water.

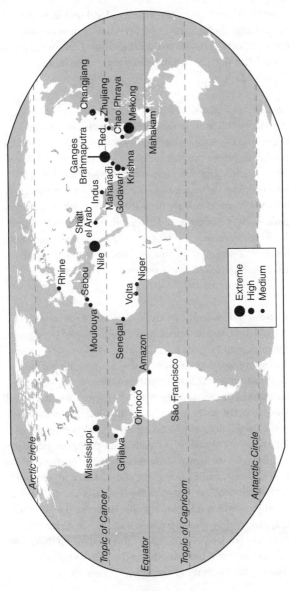

22. Areas most at risk from sea-level rise

In the 1980s, 100,000 tube wells and 20,000 deep wells were sunk, increasing the fresh-water extraction six-fold. Both these projects are essential to improving the quality of life for people in this region, but have produced a subsidence rate of up to 2.5 cm per year, one of the highest rates in the world. Using estimates of subsidence rate and global warming sea-level rise, the World Bank has estimated that by the end of the 21st century, the relative sea level in Bangladesh could rise by as much as 1.8 m. In a worst-case scenario, they estimated that this would result in a loss of up to 16 per cent of land, supporting 13 per cent of the population, and producing 12 per cent of the current GDP. Unfortunately, this scenario does not take any account of the devastation of the mangrove forest and the associated fisheries. Moreover, increased landward intrusions of salt water would further damage water quality and agriculture.

Another example of a threatened coastline is the Nile delta, which is one of the oldest intensely cultivated areas on Earth. It is very heavily populated, with population densities up to 1,600 inhabitants per km^2. Deserts surround the low-lying, fertile floodplains. Only 2.5 per cent of Egypt's land area, the Nile delta and the Nile valley, is suitable for intensive agriculture. Most of a 50 km wide land strip along the coast is less than 2 metres above sea level and is only protected from flooding by a 1–10 km wide coastal sand belt, shaped by discharge of the Rosetta and Damietta branches of the Nile. Erosion of the protective sand belt is a serious problem and has accelerated since the construction of the Aswan dam in the south of Egypt. A rising sea level would destroy weak parts of the sand belt, which are essential for the protection of lagoons and the low-lying reclaimed lands. These impacts could be very damaging. About one-third of Egypt's fish catches are made in the lagoons, and sea-level rise would change the water quality and affect most fresh-water fish; valuable agricultural land would be inundated; vital, low-lying installations in Alexandria and Port Said would be threatened; recreational tourism beach facilities would be endangered; and essential

groundwater would be salinated. Many of these effects are preventable, as dikes and protective measures would stop the worst flooding up to a 50 cm sea-level rise—although there may still be considerable groundwater salination and the impact of increasing wave action could be serious.

The most important influence on the impact of sea-level rise on coastal regions is the rate of change. At the moment, the predicted rise of about 50 cm in the next 100 years can be dealt with if there is the economic foresight to plan for the protection and adaptation of coastal regions. This then comes back to the development of regional economies and the availability of resources to implement appropriate changes. If sea level rises by over 1 m in the next 100 years, which is thought to be unlikely according to the IPCC, then humanity would doubtless have major problems adapting to it.

Storms and floods

Storms and floods are major natural hazards. Over the last decade they have been responsible for three-quarters of the global insured losses, and over half the fatalities and economic loses from natural catastrophes. It is, therefore, essential we know what is likely to happen in the future. There is evidence that the temperate regions, particularly in the Northern Hemisphere, have become more stormy over the last 50 years. In particular the climate models suggest that the proportion of rainfall occurring as heavy rainfall has and will continue to increase, as will the year-to-year variability. This will increase the frequency and magnitude of flooding events.

Two-fifths of the world's population lives under the monsoon belt, which brings life-giving rains. Monsoons are driven by the temperature contrast between continents and oceans. For example, moisture-laden surface air blows from the Indian Ocean to the Asian continent and from the Atlantic Ocean into West Africa during Northern Hemisphere summers, when the

land masses become much warmer than the adjacent ocean. In winter, the continents become colder than the adjacent oceans and high pressure develops at the surface, causing surface winds to blow towards the ocean. Climate models indicate an increase in the strength of the summer monsoons as a result of global warming over the next 100 years. There are three reasons to support why this should occur: (1) global warming will cause temperatures on continents to rise higher than those of the ocean in summer and this is the primary driving force of the monsoon system; (2) decreased snow cover in Tibet, which is to be expected in a warmer world, will increase this temperature difference between land and sea, increasing the strength of the Asian summer; (3) a warmer climate means the air can hold more water vapour, so the monsoon winds will be able to carry more moisture. For the Asian summer monsoon, this could mean an increase of 10–20 per cent in average rainfall, with an inter-annual variability of 25–100 per cent and a dramatic increase in the number of days with heavy rain. The most worrying model finding is the predicted increase in rain variability between years, which could double, making it very difficult to predict how much rainfall will occur each year—essential knowledge for farmers.

One of the more contentious areas of climate change science is the study and predictions of future tropical cyclones, or hurricanes as they are better known. Professors Kerry Emanuel (Massachusetts Institute of Technology) and Peter Webster (Georgia Institute of Technology in Atlanta) and colleagues using different methods have demonstrated that the number and intensity of hurricanes have increased over the last three decades in the North Atlantic and Western Pacific Oceans. This is because the number and intensity of hurricanes are directly linked to the sea-surface temperature. As hurricanes can only start to form if the sea-surface temperature is above 26°C it would seem sensible that in a warmer world there would be more hurricanes. Yet the formation of hurricanes is much rarer than the opportunities for them to occur. Only 10 per cent of centres of falling pressure over the

tropical oceans give rise to fully fledged hurricanes. Other considerations, such as wind shear to start the rising air spinning, need to be considered when understanding the genesis of tropical storms. In a year of high incident, perhaps a maximum of 50 tropical storms will develop to hurricane levels. Predicting the level of a disaster is also difficult as the number of hurricanes does not matter, it is whether they make landfall. For example, 1992 was a very quiet year for hurricanes in the North Atlantic Ocean. However, in August, one of the few hurricanes that year, Hurricane Andrew, hit the USA just south of Miami and caused damage estimated at $26 billion. Hurricane Andrew also demonstrates that where a storm hits is equally important: if the hurricane had hit just 20 miles (~32 km) further north it would have hit the densely populated area of Miami City and the damage bill would have more than doubled.

In terms of where hurricanes hit in developed countries, the major effect is usually economic loss, while in developing countries the main effect is loss of life. For example, Hurricane Katrina, which hit New Orleans in 2005, caused over 1,800 deaths; Hurricane Mitch, which hit Central America, killed at least 25,000 people. Hurricane Katrina was not the worst storm that has hit the USA; a storm that hit Miami in 1926 was 150 per cent larger but did little damage because Miami Beach was yet to be developed. In the USA, coastal population has doubled in the last ten years. So in terms of climate change, mitigation policies will have little effect on the costs in the developed world, while adaptation of coastal regions will be essential. But in the developing world mitigation would have a large impact in reducing the total loss of life and preventing regional economic melt-down. For example, the immediate economic impact of Hurricane Katrina was over $150 billion, but its subsequent effect on the US economy was to boost it slightly, by 1 per cent, that year due to the $105 billion injected by the Bush administration to help the reconstruction of the region. Compare this with Hurricane Mitch, which set back the economy of Central America by about a decade.

As can been seen from the case of Hurricanes Katrina and Mitch, storms and floods have the ability to destroy major cities and towns. In most cases it is the flooding that causes the worst disasters. Many major cities around the world are vulnerable to flooding because they were built close to rivers or the coast in order to facilitate trade via the oceans. London is one such city. At the moment, London is protected from flooding by the Thames Barrier. The Thames Barrier was built in response to the catastrophic floods of 1953 and was finally ready for use in 1982 (it was officially opened on 8 May 1984). The Thames Barrier protects 150 km^2 of London and property worth at least £1 trillion. Because of the foresight of previous scientific advisors to the UK government, it was built to withstand a one-in-2,000-year flood. With the increased sea level due to climate change, this risk by 2030 will increase to a one-in-1,000-year event. For example, between 1982 and 2001 it was closed 63 times. In the winter 2000/1, it was closed 24 times. In 2003, the barrier was closed for 14 consecutive tides, and in November 2007 it was closed twice for a storm surge the same size as the one that occurred in 1953. In the winter of 2013/14 it was closed 50 times which is a quarter of all the times it has been closed. At the moment, the UK economy is the sixth largest in the world, approximately £1.4 trillion per year generated through London. London is also one of the three main centres, along with New York and Tokyo, for 24-hour share-trading. If London were disabled by a major flood, then not only would this hit the economy of the UK, but potentially it could disrupt global trade and precipitate a global recession similar to that started in 2008. Hence the UK Environment Agency has put in place plans to guard against a significant sea-level rise in the future, including plans for a new barrier between Essex and Kent to guard against a possible 4 m rise in sea level.

Heat waves and droughts

As global temperatures increase, heat waves will increase. As precipitation becomes more variable and concentrated into more

intense rainfall events, so drought will increase. Heat waves and droughts, however, are relative terms, as it depends where they occur and if a region already has adaptations in place. For example heat-related death in London start at 23°C, while in Athens deaths start at 26°C. Like storms and floods, heat waves and droughts are major killers. The 2003 heat wave in Europe killed an estimated 70,000 people. Hardest hit was France with 14,800 deaths in the first three weeks of August and deaths in Paris increasing by 140 per cent. The people who die in heat waves are usually the elderly—medics call this 'harvesting', as these people were taken earlier than they would have normally died. The element that tends to kill the elderly is sustained night-time temperatures because when asleep they are unable to regulate their body temperature. After the 2003 heat wave it was realized that many of these deaths were due to the very weak public health response in France. As a result there were wide-sweeping policy changes including monitoring temperatures nationally to better predict heat waves and aid emergency preparations, improved building design and air conditioning for hospitals and retirement homes, increased training for health professionals, an emphasis on responsible media coverage with health recommendations regularly broadcast, and planned regular visits to the most vulnerable members of the population. A second heat wave occurred in 2006, and these new policies have been shown to have prevented 4,500 deaths. One of the reasons it is so difficult to understand the impacts of climate change is because people and societies can adapt to new conditions very quickly. Figure 23 shows the 2003 European heat wave in the context of summer temperatures over the last 100 years and those predicted for the next 100 years. What is clear is that the temperature of the 2003 heat wave could be the average summer temperature in 2050.

Droughts are also a major killer that should be considered. A drought happens when an area undergoes a prolonged period without sufficient water supply, whether surface or underground water. A drought can last for months or years and is usually

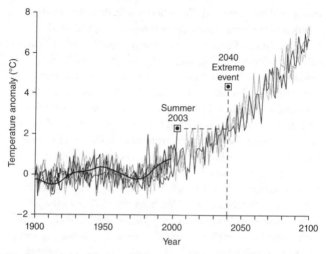

23. Comparison of the 2003 heat wave with past and future summer temperatures

caused when an area receives consistently below average rainfall. Droughts have a substantial impact on the local ecosystem and agriculture, including drops in crops growth and yield and loss of livestock. Although droughts can persist for several years, even a short, intense drought can cause significant damage and harm to the local economy. Prolonged drought has caused cause famine, mass migrations, and humanitarian crises. From a disease point of view, droughts are much worse than floods because of a lack of fresh drinking water, and stagnant pools of water. One of the major concerns with climate change is that areas vulnerable to drought will have them more frequently and areas that have never had them will start to experience them.

Human health

The potential health impacts of climate change are immense, and managing those impacts will be an enormous challenge. Climate

change will increase deaths from heat waves, droughts, storms, and floods. However, higher global temperatures will also be a challenge for many societies, particularly those that rely heavily on subsistence agriculture. As higher air temperatures will make working outside more difficult and increase the likelihood of hyperthermia. This will impact on the health of anyone who has to work outside regularly, including construction and farm workers. Conversely, it has been suggested that the death rate may drop in some countries: since many elderly people die from cold weather, warmer winters would reduce this cause of death. However, this view has been contested as recent research has shown that better housing, improved health care, higher incomes, and greater awareness of the risks of cold have been responsible for the reduction of excess winter deaths in the UK since 1950. Hence in many societies adaptation to cold climate and improved protection for the most vulnerable members of society means that warmer winters will have little or no effect in reducing the death rate.

The 2009 the University College London's *Lancet* report 'Managing the Health Effects of Climate Change' identified two major areas that could affect the health of billions of people: water and food. The most important threat to human health is lack of access to fresh drinking water. At present there are still one billion people who do not have regular access to clean, safe drinking water. Not only does the lack of water cause major health problems from dehydration, but a large number of diseases and parasites are present in dirty water. The rising worldwide human population, particularly those concentrated in urban areas, is putting a great strain on water resources. The impacts of climate change—including changes in temperature, precipitation, and sea levels—are expected to have varying consequences for the availability of fresh water around the world. For example, changes in river run-off will affect the yields of rivers and reservoirs, and thus the recharging of groundwater supplies. An increase in the rate of evaporation will also affect water supplies and contribute to the salination of irrigated agricultural lands. Rising sea levels

may result in saline intrusion in coastal aquifers. Currently, approximately 1.7 billion people, one-quarter of the world's population, live in countries that are water-stressed. By 2030 it is predicted there will be a 30 per cent increase in the demand for fresh water due to increasing access to safe drinking water, and the 50 per cent increase in demand for food and energy, both of which require large quantities of water.

Climate change is likely to have the greatest impact in countries with a high ratio of relative use to available supply. Regions with abundant water supplies will get more than they want with increased flooding. As suggested above, computer models predict much heavier rains and thus major flood problems for Europe, whilst, paradoxically, countries that currently have little water (e.g. those relying on desalination) may be relatively unaffected. It will be countries in between, which have no history or infrastructure for dealing with water shortages, that will be the most affected. In central Asia, North Africa, and southern Africa, there will be even less rainfall and water quality will become increasingly degraded through higher temperatures and pollutant run-off. Add to this the predicted increased year-to-year variability in rainfall, and droughts will become more common. Hence it is those countries that have been identified as most at risk which need to start planning now to conserve their water supplies and/or deal with the increased risks of flooding, because it is the lack of infrastructure to deal with drought and floods rather than the lack or abundance of water which causes the threat to human health.

Human health is threatened by the lack of access to affordable basic food. Future changes in temperatures, precipitation, and length of growing season will all affect the production of food and other agricultural goods. Extreme weather events must also be considered. For example the 2010 Russian heat wave led to severe droughts that reduced grain production so much that Russia banned its export to ensure there was enough for its own country.

With an increasingly globalized economy very few countries are self-sufficient in basic food and hence food imports are very important. However access to basic food is also about cost, and recent increases in oil prices, increased meat demand, and the increased production of biofuels has contributed to an increase of 80 per cent in food prices since 2004. The increase in meat eating in developing countries such as India and China is an important forcing factor because beef cattle require 8 kg of grain or meal for every kg of flesh they produce. The cost of food is also influenced by the world commodity markets. In 2008–9 there was a 60 per cent rise in the price of food and in 2011–12 there was a 40 per cent jump in price (see Figure 24). The New England Institute of Complex Study's research into these price rises shows that the underlying trend in rising prices is due to the increased price of oil, increased overall demand in oil, biofuel production, and natural disasters, but that the price spikes were due to food speculation on the global markets. So the inability of many people to afford basic food, leading to their malnutrition and starvation,

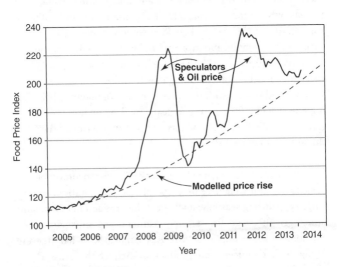

24. **Food prices. 2004–13**

can be linked directly to the speculation on food prices on the global markets in London, New York, and Tokyo.

Another possible future threat to human health is the transmission of infectious diseases, which is directly affected by climatic factors. Climate change will particularly influence vector-borne diseases, that is diseases that are carried by another organism, for example malaria, which is carried by mosquitoes. Infective agents and their vector organisms are sensitive to factors such as temperature, surface-water temperature, humidity, wind, soil moisture, and changes in forest distribution. It is, therefore, projected that climate change and altered weather patterns would affect the range (both altitude and latitude), intensity, and seasonality of many vector-borne and other infectious diseases. For example, there is a strong correlation between increased sea-surface temperature and sea level, and the annual severity of the cholera epidemics in Bangladesh. With predicted future climate change and consequent rise in Bangladesh's relative sea level, cholera epidemics could become devastating.

In general, then, increased warmth and moisture caused by climate change will enhance transmission of diseases. But while the potential transmission of many of these diseases increases in response to climate change, we should remember that our capacity to control the diseases will also change. New or improved vaccination can be expected; some vector species can be constrained by use of pesticides. Nevertheless, there are uncertainties and risks here, too: for example, long-term pesticide use encourages the breeding of resistant strains, while killing many natural predators of pests.

The most important vector-borne disease is malaria, with currently 500 million infected people worldwide. *Plasmodium vivax*, which is carried by the *Anopheles* mosquito, is the organism that causes malaria. The main climate factors that have a bearing on the malarial transmission potential of the mosquito population

are temperature and precipitation. Assessments of the potential impact of global climate change on the incidence of malaria suggest a widespread increase of risk because of the expansion of the areas suitable for malaria transmission. Mathematical models mapping out the suitable temperature zones for mosquitoes suggest that by the 2080s the potential exposure of people could increase by 2–4 per cent (260–320 million people). The predicted increase is most pronounced at the borders of endemic malarial areas and at higher altitudes within malarial areas. The changes in malaria risk must be interpreted on the basis of local environmental conditions, the effects of socioeconomic development, and malaria control programmes or capabilities. The incidence of infection is most sensitive to climate changes in areas of South-East Asia, South America, and parts of Africa. Climate change will also provide excellent conditions for *Anopheles* mosquitoes to breed in southern England, continental Europe, and the northern USA.

It should, however, be noted that the occurrence of most tropical diseases is related to development. As recently as the 1940s, malaria was endemic in Finland, Poland, Russia, and 36 states in the USA including Washington, Oregon, Idaho, Montana, North Dakota, New York, Pennsylvania, and New Jersey. So although global warming has the potential to increase the range of many of these tropical diseases, the experience of Europe and the USA suggests that combating malaria is strongly linked to development and resources: development to ensure efficient monitoring of the disease and resources to secure a strong effort to eradicate the mosquitoes and their breeding grounds.

Biodiversity

The current loss of biodiversity around the world is due to human activity including deforestation, agriculture, urbanization, and mineral exploitation. Extinction rates are currently 100–1,000 times higher than the background natural rate and climate change

will exacerbate this decline. The IPCC 2014 impact report lists the following species as those most at risk from climate change: the mountain gorilla in Africa; amphibians that live in the cloud forests of the neotropics; the spectacled bear of the Andes; forest birds of Tanzania; the 'resplendent quetzal' bird in Central America; the Bengal tiger, and other species found only in the Sundarban wetlands; rainfall-sensitive plants found only in the Cape Floral Kingdom of South Africa; and polar bears and penguins near the poles. The primary reason for the threat to these species is that they are unable to migrate in response to climate change because of their particular geographical location or the encroachment of human activity, particularly farming and urbanization. An example of the former is the cloud forests of the neotropics: as climate changes, this particular climatic zone will migrate up the mountainside until the point where there is no more mountain left.

One example of an ecosystem under threat is the coral reefs. Coral reefs are a valuable economic resource for fisheries, recreation, tourism, and coastal protection. Some estimate that the global cost of losing the coral reefs runs into hundreds of billions of dollars each year. In addition, reefs are one of the largest global stores of marine biodiversity. The last few years have seen unprecedented declines in the health of coral reefs. In 1998, El Niño was associated with record sea-surface temperatures and associated coral bleaching, which is when the coral expels the algae that live within it that are necessary to its survival (see Box 2). In some regions, as much as 70 per cent of the coral may have died in a single season. There has also been an upsurge in the variety, incidence, and virulence of coral disease in recent years, with major die-offs in Florida and much of the Caribbean region. In addition, increasing atmospheric carbon dioxide concentrations could decrease the calcification rates of the reef-building corals, resulting in weaker skeletons, reduced growth rates, and increased vulnerability to erosion. Model results suggest these effects would be most severe at the current margins of coral reef distribution.

Box 2 El Niño–Southern Oscillation

One of the most important and mysterious elements in global climate is the periodic switching of the direction and intensity of ocean currents and winds in the Pacific. Originally known as El Niño ('Christ child' in Spanish), as it usually appears at Christmas, and now more normally known as ENSO (El Niño–Southern Oscillation), this phenomenon typically occurs every three to seven years. It may last from several months to more than a year. The 1997–8 El Niño conditions were the strongest on record and caused droughts in the southern USA, East Africa, northern India, north-east Brazil, and Australia. In Indonesia, forest fires burned out of control in the very dry conditions. In California, parts of South America, Sri Lanka, and East Central Africa, there were torrential rains and terrible floods.

ENSO is an oscillation between three climates: 'normal' conditions, La Niña, and 'El Niño'. El Niño conditions have been linked to changes in the monsoon, storm patterns, and occurrence of droughts all over the world. The state of the ENSO has also been linked to the position and occurrence of hurricanes in the Atlantic. For example, it is thought that the poor prediction of where Hurricane Mitch made landfall was because the ENSO conditions were not considered and the strong trade winds helped drag the storm south across Central America instead of west as predicted.

Predicting El Niño events is very difficult but getting steadily easier. For example, there is now a large network of both ocean and satellite monitoring systems over the Pacific Ocean, primarily aimed at recording sea-surface temperature, which is the major indicator of the state of the ENSO. By using this climatic data in both computer circulation models and statistical models, predictions are made of the likelihood of an El Niño or La Niña event. We are really still in the infancy stage of developing our understanding and predictive capabilities of the ENSO phenomenon.

There is also considerable debate over whether ENSO has been affected by climate change. The El Niño conditions generally occur every three to seven years and in the last 30 years this has continued by with no discernable pattern. As El Niño conditions occurred in 1987–8, 1991–2, 1994–5, 1997–8, 2003–4, and 2010–11 representing gaps of four, three, three, six, and seven years, respectively. While La Niña events occurred in 1989–90, 1997–8, 1999–2000, 2000–1, 2008–9, 2011–12, and 2012–13, representing gaps of eight, two, zero, eight, three, and, again, zero years, respectively. Reconstruction of past climate using coral reefs in the western Pacific shows sea-surface temperature variations dating back 150 years, well beyond our historical records. The sea-surface temperature shows the shifts in ocean current, which accompany shifts in the ENSO and reveal that there have been two major changes in the frequency and intensity of El Niño events. First was a shift at the beginning of the 20th century from a 10–15-year cycle to a 3–5-year cycle. The second was a sharp threshold in 1976 when a marked shift to more intense and even more frequent El Niño events occurred. Some climate modelling results suggest that there could be a future 'heightened' state of El Niño which would permanently shift weather patterns. For example, it seems that the drought region in the USA could be shifting eastward. However, predicting an El Niño event six months from now is hard enough, without trying to assess whether or not ENSO is going to become more extreme over the next 100 years. Most computer models of ENSO in the future are inconclusive; some have found an increase and others have found no change. This is, therefore, one part of the climate system which we do not know how global warming will affect. Not only does ENSO have a direct impact on global climate, but it also affects the numbers, intensity, and pathways of hurricanes and cyclones, and the strength and timing of the Asian monsoon. Hence, when discussing the potential impacts of climate change, one of the largest unknowns is the variation of ENSO and its knock-on effects on the rest of the global climate system.

On a more theoretical note, a recent study by Chris Thomas and colleagues (published in *Nature* in 2004) investigated the possible increase in the likely extinction rate over the next 50 years in key regions such as Mexico, Amazonia, and Australia. The theoretical models suggest that by 2050 the climatic changes predicted by the IPCC would commit 18 per cent (warming of 0.8–1.7°C), 24 per cent (1.8–2.0°C), and 35 per cent (above 2.0°C) of the species studied to extinction in these regions. That means one-quarter of all species in these regions may be committed to extinct by the middle of this century. This study has been criticized as there are many assumptions in their models which may or may not be true; for example, they assume we know the full climatic range in which each species can persist and the precise relationship between shrinking habitat and extinction rates. So these results should be seen only as the likely direction of extinction rates, not necessarily the exact magnitude. However, these predictions do represent a huge future threat to regional and global biodiversity and illustrate the sensitivity of biological systems to the amount and rate of warming that will occur in the future.

Agriculture

One of the major worries concerning future climate change is the effect it will have on agriculture, both globally and regionally. The main question is whether the world can feed itself with an extra two billion people on the planet by 2050 and a rapidly changing climate. Figure 25 shows the changes in cereal grain yields that could occur with a business-as-usual carbon emissions by 2050. It is clear that as some countries', particularly in the higher latitudes, agricultural productivity improves, it will be the poorest countries in the Tropics that will suffer the most. This prediction assumes that farmers will not adapt to changing climate as that would boost or at least maintain agricultural production in many regions. For example, farmers can vary the planting time and/or switch to a different variety of the same plant to respond to changing conditions. Models suggest that,

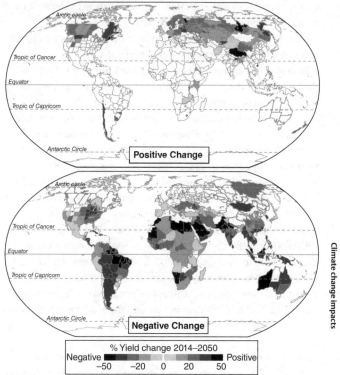

25. Changes in cereal grain yields by 2050

with reasonable assumptions made on a worldwide scale, the
change should be small or moderate. But this does not mean
the amount of cereal produced worldwide will be the same or
lower in 2060 compared with today. Since 1960, world grain
production has doubled and is predicted to continue to rise at
a similar rate.

The general global trend, however, masks the huge changes that
will occur regionally, with both winners and losers—the losers
being the poorest countries, of course, as they are least able to

91

adapt. Also, the results of all these studies are heavily dependent on the assumed trade models and market forces used, as, unfortunately, agricultural production in the world has very little to do with feeding the world's population and much more to do with trade and economics. This is why the European Union has stockpiles of food, while many underdeveloped countries export cash crops (such as sugar, cocoa, coffee, tea, and rubber) but cannot adequately feed their own populations. A classic example is the West African state of Benin, where cotton farmers can obtain cotton yields four to eight times per hectare greater than their US competitors in Texas. However, because the USA subsidizes its farmers, this means that US cotton is cheaper than that coming from Benin. Currently, US cotton farmers receive over $4 billion in subsidies, almost twice the total GDP of Benin. In 2002 Brazil filed a case with the World Trade Organization (WTO) against the USA for unfair subsidises and distortion of trade. They won their case in 2005, however, seven years later, the USA is still discussing what changes should be made to their farming subsidies. So even if climate change makes Texan cotton yields even lower, it still does not change the biased market forces still illegally in operation.

So markets can reinforce the difference between agricultural impacts in developed and developing countries and, depending on the trade model used, agricultural exporters may gain in monetary terms even though the supplies fall, because when a product becomes scarce, the price rises. The other completely unknown factor is the extent to which a country's agriculture can be adapted. For example, in climate change models it is assumed that production levels in developing countries will fall to a greater degree than those in developed countries because their estimated capability to adapt is less than that of developed countries. But this is just another assumption that has no analogue in the past, and as these effects on agriculture will occur over the next century, many developing countries may catch up with the developed world in terms of adaptability.

One example of the real regional problems that climate change could cause is the case of coffee growing in Uganda. Here, the total area suitable for growing Robusta coffee would be dramatically reduced, to 10 per cent of the present area, by a temperature increase of 2°C. Only higher areas of land would remain suitable for coffee growing, the rest would become too hot. But no one can tell whether these remaining areas would make more or less money for the country because if other coffee growing areas around the world are similarly affected, the price of coffee beans will increase due to scarcity. This demonstrates the vulnerability to the effects of global warming of many developing countries, whose economies often rely heavily on one or two agricultural products, as it is very difficult to predict the changes that global warming will cause in terms of crop yield and its cash equivalent. Hence one major adaptation to global warming should be the broadening of the economic and agricultural base of the most threatened countries. This, of course, is much harder to accomplish in practice than on paper, and it is clear that the EU and US agricultural subsidies and the current one-sided world trade agreements have a greater effect on global agricultural production and the ability of countries to feed themselves than climate change will ever have. Solutions look even further away with the failure of the WTO negotiations that have collapsed again.

Ocean acidification

Direct measurements of the ocean's chemistry have shown that it has become more acidic (see Figure 26), which is shown by its lower pH. This is because carbon dioxide in the atmosphere dissolves in the water of the surface ocean. This is controlled by two main factors: the amount of carbon dioxide in the atmosphere and the temperature of the ocean. The oceans have already absorbed about a third of the carbon dioxide resulting from human activities, which has lead to a steady decrease in ocean pH levels. With increasing atmospheric carbon dioxide in the future the amount of dissolved carbon dioxide in the ocean will continue

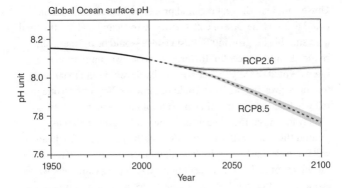

Global Ocean surface pH

26. Ocean acidification

to increase. Some marine organisms, such as corals, foraminifera, coccoliths, and shellfish, have shells composed of calcium carbonate which dissolves more readily in acid. Laboratory and field experiments show that under high carbon dioxide the more acidic waters cause some marine species to develop misshapen shells and have lower growth rates, although the effect varies among species. Acidification also alters the cycling of nutrients and many other elements and compounds in the ocean, and it is likely to shift the competitive advantage among species, and have impacts on marine ecosystems and the food web. This is a major worry as fishing is still a major source food, with about 95 million tonnes caught by commercial fishing and another 50 million tonnes produced by fish farms per year.

Summary

The impacts of climate change will increase significantly as the temperature of the planet rises. Climate change will affect the return period and severity of floods, droughts, heat waves, and storms. Coastal cities and towns will be especially vulnerable as sea-level rise will worsen the effects of floods and storm surges. Water and food security as well as public health will become the

Table 4. Impacts of climate change

Temperature rise above pre-industrial levels	Major potential impacts
1°–2°C	• Major impacts on vulnerable ecosystems and species (such as polar regions, wetlands, and cloud forests).
	• Increase in extreme weather events and spread of infectious disease.
2°–3°C	• Major loss of coral reef ecosystem.
	• Major impacts on all ecosystems and species.
	• Large impacts on agriculture, water resources, and human health.
	• Significant increase in extreme weather events.
	• Terrestrial carbon sink becomes a source, accelerating climate change.
3°–4°C	• Major species extinction.
	• Food and water security become major issues.
	• Significant impacts on human health via lack of food and clean water disproportionately effecting the poor.
	• Environmental forced mass migration increases.
	• Ocean carbon sink greatly reduces accelerating carbon accumulation in the atmosphere.

(continued)

Table 4. Continued

Temperature rise above pre-industrial levels	Major potential impacts
	• Extreme weather events are over 10 times more common than in 2010.
>4°C	• Western Antarctic and Greenland ice sheets melting accelerates causing great rises in global sea level.
	• Environmental forced mass migration accelerates and there could be increase conflict over resources.
4°–5°C	• Fifth of world population effected by flooding.
	• Over 3 billion people suffer from water scarcity.
	• Food yields fall everywhere and global production plummets leading to wide spread malnutrition and starvation.
	• Significant increase in human deaths due to malnutrition, disease, flooding, and extreme weather events.
5°–6°C or higher	• Don't go there.

most important problems facing all countries. Climate change threatens global biodiversity and the wellbeing of billions of people.

In Table 4, there is a summary of the effects of climate change by the rise in global temperature. Because of the slowness of the political process many scientists and social scientists have started to think about what a 4°C world would be like. Very few of us discuss the possibilities of a 5°–6°C world, even the IPCC has avoided discussing a world that hot. It is, however, important for

us to have an understanding of what this sort of climate change would mean to the planet to ensure we never go there. With sustained global temperatures of 5°–6°C above present-day levels, both Greenland and the Western Antarctic ice sheet will be committed to full melting possibly by the middle to end of the next century. If these two ice sheets completely melt, sea level would rise by 13 m. At the moment, the UK Environment Agency has plans to deal with a 4 m rise in sea level, which entail a huge barrier across the mouth of the River Thames, stretching from Essex to Kent. However, a sea-level rise of 13 m would mean the flooding and permanent abandonment of nearly all lower lying coastal and urban river areas. At the moment, one-third of the world's population lives within 60 miles (~96 km) of a shoreline and 13 of the world's 20 largest cities are located on a coast. This means billions of people could be displaced leading to mass migration. Depending on whether Greenland or the Western Antarctic ice sheet melted first the North Atlantic Ocean circulation could collapse, creating seasonal extremes for Western Europe, with very cold winters followed by heat waves every summer. At least three billion people in the world would become water-stressed. Agricultural production would start to fail, and billions of people would face starvation. Water and food security would then become issues of conflict between countries, with some experts predicting 'eco-wars'. Public health systems around the world could collapse, unable to cope with the demands. And global biodiversity would be devastated. Let's not go there.

Chapter 6
Climate surprises

We are currently changing the composition of the atmosphere beyond anything that has been experienced over the last few millions of years. We are headed for unknown territory and therefore scientific uncertainty can be large. However, we know from the study of past climates that the climate system can switch into a new state very quickly once a threshold has been past. For example, ice-core records suggest that half the warming in Greenland since the last ice age was achieved in only a few decades. This chapter examines the possibility that there are thresholds or tipping points in the climate system that may occur as we warm the planet. Figure 27 shows the main tipping points which scientists have been concerned about over the last two decades. Irreversible melting of the Greenland and/or Western Antarctic ice sheet, slowing down of the North Atlantic deep ocean circulation, gas hydrates, and the Amazon rainforest dieback will all be discussed.

Thresholds and tipping points

The relationship between a climate forcing such as GHGs and the climate response is complicated. In an ideal world it would be a simple relationship with little or no delay, but we already know that there is inertia in the climate system so that it responds to

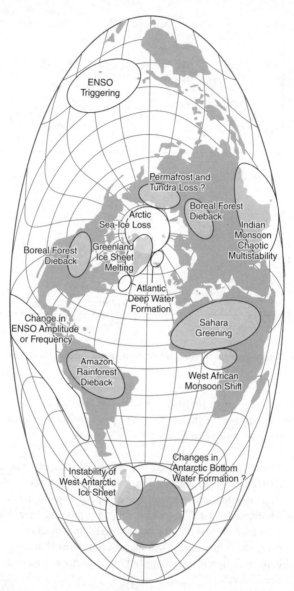

27. Climate change tipping points

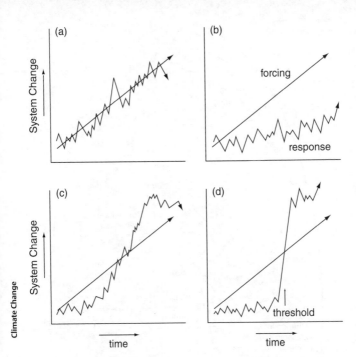

28. Greenhouse gas forcing and the response of the climate system

GHG forcing with a 10-20 year delay. So we can examine the way different parts of the climate system respond to climate change with four scenarios (see Figure 28).

(a) *Linear but delayed response (Figure 28a)*. In this case, the GHGs increase produces a delayed but direct response in the climate system whose magnitude is in proportion to the additional forcing. This can be equated to pushing a car along a flat road. At first nothing happens as friction must be overcome before the car will start moving. Once that has happened most of the energy put into pushing is used to move the car forward.

(b) *Muted or limited response (Figure 28b)*. In this case, the GHG forcing may be strong, but the part of the climate system is in some way buffered and therefore gives very little response. An example of this is the East Antarctic ice sheet that has been stable at much warmer temperatures than today. This is the 'pushing the car up the hill' analogy: you can spend as much energy as you like trying to push the car, but it will not move very far.

(c) *Delayed and non-linear response (Figure 28c)*. In this case, the climate system may have an initial slow response to the GHG forcing but then responds in a non-linear way. This is a real possibility when it comes to climate change if we have underestimated the positive feedback in the system. This scenario can be equated to the car on the top of a hill: it takes some effort and thus time to push the car to the edge of the hill; this is the buffering effect. Once the car has reached the edge, it takes very little effort to push the car over, and then it accelerates down the hill with or without help. Once it reaches the bottom, the car then continues for some time, which is the overshoot, and then slows down of its own accord and settles into a new state.

(d) *Threshold response (Figure 28d)*. In this case, initially, there is very little response in that part of the climate system to the GHG forcing; however, all the response takes place in a very short period of time in one large step or threshold. In many cases, the response may be much larger than one would expect from the size of the forcing and this can be referred to as a 'response overshoot'. An example of this could be the Greenland ice sheet that has started to melt but that could sudden accelerate causing a catastrophic collapse. This scenario equates to the bus hanging off the cliff at the end of the original film *The Italian Job*; as long as there are only very small changes, nothing happens at all. However, a critical point (in this case weight) is reached and the bus (and the gold) plunges off the cliff into the ravine below.

Though these are purely theoretical models of how the global climate system can respond, they are important to keep in mind when reviewing the possible scenarios for future climate change. An added complication when assessing climate change is the possibility that climate thresholds contain bifurcations. This means the forcing required to push the system one way across the threshold is different from the reverse. This implies that once a climate threshold has occurred, it is a lot more difficult to reverse it and in some cases it may be irreversible.

The term 'tipping points' is used a lot in climate change research and discussions. However care must be taken as there are two usage of this word. First are the climate tipping points, which are the large scale shifts in the climate system, such as irreversible melting of ice sheets or the release are huge stores if methane below the oceans. The others are societal tipping points when climate change has a major effect on a region or a particular country. For example a 200-mile (~322-km) shift northward of the South-East Asian monsoonal rainfall belt in climatologically terms is a small shift and is not a fundamental climate tipping point. But for the countries where the rains no longer fall or those where it does for the first time this is a major climate tipping point because their weather may have been permanently altered.

Melting icesheets

The IPCC sea-level rise projections for 2100 with no curbs to carbon emissions are between 57 cm and 98 cm. The largest uncertainty within these estimates is the contribution that the melting of Greenland and Antarctica will make by the end of the century. At the moment it is estimated that Greenland is losing over 200 gigatonnes of ice per year, a six-fold increase since the early 1990s. While Antarctica is losing about 150 gigatonnes of ice per year, a five-fold increase since the early 1990s. Most of this loss is from the northern Antarctic Peninsula and the Amundsen

sea sector of West Antarctica. Greenland and Antarctica constitute one of the most worrying potential climate surprises. If the large ice sheets there completely melted, their contribution to global sea-level rise would be as follows: Greenland, about 7 m; West Antarctic ice sheet, about 8.5 m; East Antarctic ice sheet, about 65 m; compared with just 0.3 m if all the mountain glaciers melted. Palaeoclimate data show that the huge East Antarctic ice sheet developed 35 million years ago due to the progressive tectonic isolation of Antarctica and that it has in fact remained stable in much warmer climates. So climate scientists have a very high degree of confidence that the East Antarctic ice sheet will remain stable in this century. In fact there is evidence that the warmer wetter climate is allowing a small increase in snow accumulation on the ice sheet. However, scientists are now very worried that the melting of Greenland or the West Antarctic could significantly accelerate in next 100 years. Even if we have already started the processes of melting the whole of these ice sheets, there is of course a physical constraint to the speed that the ice can melt. This is due to the time it takes for heat to penetrate into the ice sheets. Imagine dropping an ice cube into a hot cup of coffee you know it will melt entirely, but it takes time for the heat to penetrate into the middle of the ice cube. Also most of the ice from ice sheets goes through ice streams, which have a limit on how much ice they can mobilize. The worst case scenario according to leading glaciologists is that these ice sheets could add between 1 m to 1.5 m to the sea level by the end of the century, which would threaten many coastal populations around the world. There is also scientific debate about what happens to both the Greenland and Antarctic ice sheets beyond the next 100 years; even if significant melting does not occur this century, we may have started a process that causes irreversible melting during the next one. Thus our carbon emission over the next few decades could determine the long-term future of these ice sheets and the livelihoods of billions of people who live close the coast.

Deep-ocean circulation

The circulation of the ocean is one of the major controls on our global climate. In fact, the deep ocean is the only candidate for driving and sustaining internal long-term climate change (of hundreds to thousands of years) because of its volume, heat capacity, and inertia. In the North Atlantic, the north-east trending Gulf Stream carries warm and salty surface water from the Gulf of Mexico up to the Nordic seas. The increased saltiness, or salinity, in the Gulf Stream is due to the huge amount of evaporation that occurs in the Caribbean, which removes moisture from the surface waters and concentrates the salts in the sea water. As the Gulf Stream flows northward, it cools down. The combination of a high salt content and low temperature makes the surface water heavier or denser. Hence, when it reaches the relatively fresh oceans north of Iceland, the surface water has cooled sufficiently to become dense enough to sink into the deep ocean. The 'pull' exerted by the sinking of this dense water mass helps maintain the strength of the warm Gulf Stream, ensuring a current of warm tropical water flowing into the north-east Atlantic, sending mild air masses across to the European continent. It has been calculated that the Gulf Stream delivers 27,000 times the energy of all of Britain's power stations put together. If you are in any doubt about how good the Gulf Stream is for the European climate, compare the winters at the same latitude on either side of the Atlantic Ocean, for example London with Labrador, or Lisbon with New York. Or a better comparison is between Western Europe and the west coast of North America, which have a similar geographical relationship between the ocean and continent—so think of Alaska and Scotland, which are at about the same latitude.

The newly formed deep water sinks to a depth of between 2,000 m and 3,500 m in the ocean and flows southward down the Atlantic Ocean, as the North Atlantic Deep Water (NADW). In

the South Atlantic Ocean, it meets a second type of deep water, which is formed in the Southern Ocean and is called the Antarctic Bottom Water (AABW). This is formed in a different way to NADW. Antarctica is surrounded by sea ice and deep water forms in coast polynyas, or large holes in the sea ice. Out-blowing Antarctic winds push sea ice away from the continental edge to produce these holes. The winds are so cold that they super-cool the exposed surface waters. This leads to more sea-ice formation and salt rejection, producing the coldest and saltiest water in the world. AABW flows around the Antarctic and penetrates the North Atlantic, flowing under the warmer, and thus somewhat lighter, NADW (see Figure 29a). The AABW also flows into both the Indian and Pacific Oceans.

This balance between the NADW and AABW is extremely important in maintaining our present climate, as not only does it keep the Gulf Stream flowing past Europe, but it also maintains the right amount of heat exchange between the Northern and Southern Hemispheres. Scientists have shown that the circulation of deep water can be weakened or 'switched off' if there is sufficient input of fresh water to make the surface water too light to sink. This evidence has come from both computer models and the study of past climates. Scientists have coined the phrase 'dedensification' to mean the removal of density by adding fresh water and/or warming up the water, both of which prevent sea water from being dense enough to sink. As we have seen, there is already concern that global warming will cause significant melting of the polar ice caps. This will lead to more fresh water being added to the polar oceans. Climate change could, therefore, cause the collapse of NADW, and a weakening of the warm Gulf Stream (Figure 29b). This would cause much colder European winters and more severe weather. However, the influence of the warm Gulf Stream is mainly seen in the winter and has only a small effect on summer temperatures. So, if the Gulf Stream fails, global warming would still cause European summers to heat up. Europe would end up with extreme seasonal weather very similar to that of Alaska.

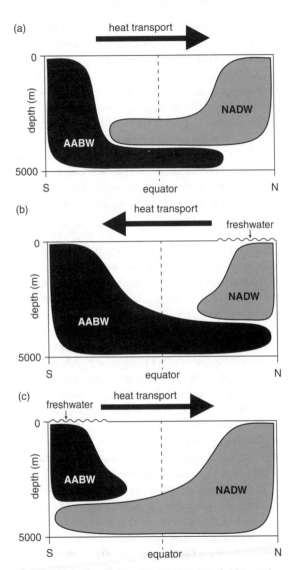

29. Deep ocean circulation changes depending on freshwater inputs
NADW = North Atlantic Deep Water; and AABW = Antarctic Bottom Water.

A counter scenario is that, if the Antarctic ice sheet starts to melt significantly before the Greenland and Arctic ice, things could be very different. If enough melt-water goes into the Southern Ocean, then AABW will be severely curtailed. Since the deep-water system is a balancing act between NADW and AABW, if AABW is reduced then the NADW will increase and expand (Figure 29c). The problem is that NADW is warmer than AABW, and because if you heat up a liquid it expands, the NADW will take up more space. So any increase in NADW could mean an increase in sea level. Computer models by Dan Seidov (now at the National Oceanic and Atmospheric Administration) and myself have suggested that a melt-water event in the Southern Ocean could cause a reduction in the AABW and the expansion of the NADW, and would result in an average sea-level increase of over 1 m. It has been over 20 years since the possibility of a catastrophic shut down of the deep ocean circulation was suggested and there has been a huge amount of work on it. The consensus in the very latest IPCC science report is that this collapse is highly unlikely in the 21st century. The evidence is from detailed monitoring of the oceans and climate models, and a greater understanding of the past changes in ocean circulation. Put simply, there will not be enough freshwater entering either the Nordic Seas or the Southern Ocean in the next 100 years or so to switch off the circulation. However, if emissions keep rising as they have over the last 10 or 20 years then by the next century when both Greenland and the Western Antarctica are committed to full melt then we could see huge amounts of freshwater entering the oceans and deep-ocean circulation will be disrupted.

Gas hydrates

Currently, below the world's oceans and permafrost is stored a large amount of carbon in the form of gas hydrates. These are a mixture of water and methane, which is solid at low temperatures and high pressures. These gas hydrates are cages of water molecules, which hold individual molecules of methane and other

gases within the cage. The methane comes from decaying organic matter found deep in ocean sediments and in soils beneath permafrost. These gas hydrate reservoirs could be unstable, as an increase in temperature or decrease in pressure would cause them to destabilize and to release the trapped methane. Climate change is warming up of both the oceans and the permafrost, threatening the stability of gas hydrates. Methane is a strong GHG, 21 times more powerful than carbon dioxide. If enough were released, it would raise global temperatures, which could led to the release even more gas hydrates—producing a runaway effect. Scientists really have no idea how much methane is stored in the gas hydrates beneath our feet: estimates are between 1,000 and 10,000 gigatonnes, which is a huge range (compared with ~800 gigatonnes of carbon currently in the atmosphere). Without a more precise estimate, it is very difficult to assess the risk posed by gas hydrates.

The reason why scientists are so worried about this issue is because there is evidence that a super greenhouse effect occurred 55 million years ago, during what is called the Palaeocene–Eocene Thermal Maximum (PETM). During this hot-house event, scientists think that up to 1,500 gigatonnes of gas hydrates may have been released. This huge injection of methane into the atmosphere accelerated the natural greenhouse effect, producing an extra 5°C of warming. There is still, however, considerable debate over the PETM. For example, was it gas hydrate or volcanic carbon dioxide release that caused the warming? The current consensus is that the ocean reserves of gas hydrate are likely to remain stable this century. This is because as ocean temperatures change, the change has to be transmitted through the gas hydrate layer, causing some of it to melt, but if this process is slow enough, the gas released might migrate up in the ocean sediment column and re-freeze at a higher level. However, if carbon emissions are not curbed then by the next century we could see this process speed up leading to the release of some of the methane stored in the deep ocean. It is

also clear that there will be some permafrost gas hydrate destabilization as climate change is having the greatest impact on high latitude temperatures, and permafrost melting has been reported in almost all areas. However, we still do not have an indication of how much methane is stored beneath the world's permafrost regions. So at the moment our best estimate suggests a global warming of 3°C could release between 35 and 940 gigatonnes of carbon, which could add between 0.02°C to 0.5°C to global temperatures.

There is another problem. If significant parts of the Greenland and Antarctic ice sheets melt, the removal of ice from the continent means that it will recover and start to move upwards. This isostatic rebound can be seen in the British Isles, which are still recovering from the last ice age, with Scotland still rising while England is sinking. This will mean that the relative sea level around the continental shelf will fall, removing the weight and thus the pressure of the seawater on the marine sediment. Pressure removal is a much more efficient way of destabilizing gas hydrates than temperature increases, and so huge amounts of methane could be released from around the Arctic and Antarctic.

There is another secondary effect of gas hydrate release: when the hydrates break down, they can do so explosively. There is clear evidence from the past that violent gas hydrate releases have caused massive slumping of the continental shelf and associated tsunamis (giant waves). Eight thousand years ago the Norwegian Storegga slide, which was the size of Wales, produced a 15 m high tsunami that wiped out many prehistoric settlements in Scotland. In modern times, we have seen the destruction caused by the 2004 Boxing Day Indian Ocean and the 2011 Japanese tsunamis. Hence we cannot rule out the possibility that climate change could lead to an increased frequency of gas hydrate generated submarine landslides, and thus tsunamis of over 15 m in height hitting our coasts. Up to now, only the countries around the Pacific rim are prepared for this type of

tsunami event—but gas hydrate generated tsunamis could occur anywhere in the ocean.

Amazon dieback

In 1542, Francisco de Orellana led the first European voyage down the Amazon River. During this intrepid voyage the expedition met a lot of resistance from the local Indians; in one particular tribe the women warriors were so fierce that they drove their male warriors in front of them with spears. Thus the river was named after the famous women warriors of the Greek myths, the Amazons. This makes Francisco de Orellana one of the unluckiest explorers of that age, as normally the river would have been named after him. The Amazon River discharges approximately 20 per cent of all fresh water carried to the oceans. The Amazon drainage basin is the world's largest, covering an area of 7,050,000 km^2, about the size of Europe. The river is a product of the Amazon monsoon, which every summer brings huge rains. This also produces the spectacular expanse of rainforest, which supports the greatest diversity and largest number of species of any area in the world. The Amazon rainforest is important when it comes to climate change as it is a huge natural store of carbon. Up until recently, it was thought that because an established rainforest such as the Amazon had reached maturity it could not take up any more carbon dioxide. Experiments in the heart of the Amazon rainforest have shown this assumption is wrong and that the Amazon rainforest might be sucking up an additional 5 tonnes of atmospheric carbon dioxide per hectare per year. Indeed a paper in *Nature* in 2014 by Dr Stephenson (US Geological Survey) and colleagues showed, for both tropical and temperate tree species, that carbon storage increases continuously with tree size. Thus old trees do not act simply as carbon reservoirs but they actively fix large amounts of carbon compared to smaller trees. At one extreme, a single big tree can add the same amount of carbon to the forest within a year as that contained in an entire mid-sized tree.

The concern about a possible Amazon rainforest dieback came from a seminal paper published in 2000 by colleagues at the UK Meteorological Office's Hadley Centre. Their climate model was the first to have vegetation-climate feedback and suggest that global warming by 2050 could have increased the winter dry season in Amazonia. For the Amazon rainforest to survive, it requires not only a large amount of rain during the wet season but a relatively short dry season so as not to dry out. According to the Hadley Centre model, climate change could cause the global climate to shift towards a more El Niño-like state with a much longer South American dry season. Kim Stanley Robinson in his novel *Forty Signs of Rain* uses the term 'Hyperniño' to refer to a new climate state. Hence the Amazon rainforest could no longer survive and would be replaced by savannah (dry grassland), which is found both to the east and south of the Amazon basin today. This replacement would occur because the extended dry periods would lead to forest fires destroying large parts of the rainforest. This is exactly what was seen during the two extreme Amazon droughts of 2005 and 2010. This also returns the carbon stored in the rainforest back into the atmosphere, accelerating climate change. The savannah would then take over those burnt areas, as it is adapted to coping with the long dry season, but savannah has a much lower carbon storage potential per km^2 than rainforest. However, this concern was premature, as other climate models could not find the same extreme response of the climate over Amazonia. Key to this was the Coupled Carbon Cycle Climate Model Intercomparison Project (C^{4}MIP), which compared global climate models that include an interactive vegetation and carbon cycle. The results of C^{4}MIP and the opinion of the latest IPCC review is that it is unlikely that a sustained dieback of the Amazon rainforest will occur this century due to climate change. However, deforestation is still occurring in the Amazon due to logging, mining, and the expansion of agriculture. Moreover, the huge droughts of 2005 and 2010 did show how vulnerable the Amazon rainforest is to extreme variations in dry-season length. While much of the

damage has started to grow back, there still is concern about long-term loss due to direct human impacts.

Summary

Until a few decades ago, it was generally thought that significant large-scale global and regional climate changes occurred gradually over many centuries or millennia, hence the climate shifts were assumed to be scarcely perceptible during a human lifetime. We now know that human-induced climate change will radically affect the planet over the next 100 years. In addition, there may be potential surprises in the global climate system, exacerbating future climate change. As discussed above, these include the possibility that Greenland and/or the Antarctic could start to irreversibly melt, raising sea level by many metres in the next century. The North Atlantic driven deep-ocean circulation could change, producing extreme seasonal weather in Europe. The Amazon rainforest could start to die back due to the combined effects of deforestation and climate change, causing the loss of huge amounts of biodiversity. Finally, there is the threat of additional methane being released from gas hydrates beneath the oceans and permafrost, which could accelerate climate change.

So, what effects could climate change have on human society? We know that abrupt past climate changes had profound effects on human history. For example, a short, cold, arid period about 4,200 years ago caused the collapse of classical civilizations around the world, including the Old Kingdom in Egypt; the Akkadian Empire in Mesopotamia; the Early Bronze Age societies of Anatolia, Greece, and Israel; the Indus Valley civilization in India; the Hilmand civilization in Afghanistan; and the Hongshan culture of China. It has also been shown that climate deterioration, particularly a succession of severe droughts in Central America during the Medieval Cold Period, prompted the collapse of the classic period of the Mayan civilization. Moreover, the rise and fall

of the Incas can be linked to alternating wet and dry periods, which favoured the coastal and highland cultures of Ecuador and Peru.

We know, however, that humans can survive a whole range of climates. The collapse of these urban civilizations was not simply about climate shifts making an area inhospitable, rather, it was that those societies were unable to adapt to the climate changes, particularly the changes in water resources. For example, for the Mayan civilization to have survived, it would have needed to recognize its vulnerability to long-term water shortages and to have developed a more flexible approach, such as finding new water sources, developing new means of conserving water, and prioritizing water use in times of shortage. Hence the next two chapters are concerned with the global human response to climate change and a discussion of potential solutions, including how to ensure our civilization becomes flexible enough to deal with the possibility of climate surprises.

Chapter 7
Politics of climate change

Introduction

The most logical approach to the climate change problem would be to significantly cut GHG emissions. At the moment, many countries are developing very rapidly and thus global emissions are expanding at a faster and faster rate. So how much do we need to cut emissions by? As we have seen in previous chapters, scientists feel that 2°C is the tipping point, when almost all people in the world become losers from climate change. So limiting climate change to 2°C seems to be the logical thing to do. Especially as the Stern Report in 2007 suggested the cost of adapting to a low-carbon economy now would be about 1–3 per cent of world GDP compared to costs of at least 20 per cent world GDP if we do nothing. To try to limit climate change to 2°C we need to understand how much carbon dioxide and other GHGs this represents. Figure 30 shows the probability of temperature changes based on different amounts of atmospheric carbon dioxide. Even at the lowest level of 450 parts per million (ppm), there is at least a 40 per cent chance that climate change will be above 2°C. Remember that at the moment we are already at 400 ppm and increasing at over 2 ppm per year so that 2°C limit could be reached in less than 25 years. We face a huge challenge if we are to contain climate change to this limit, and the only way we will do that is by an

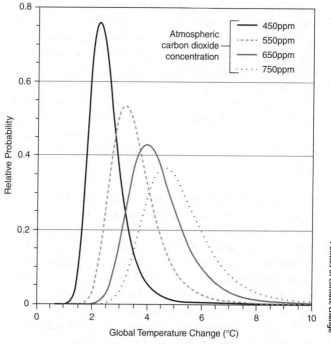

30. Predicted range of global temperature rise based on the amount of carbon dioxide in the atmosphere

international treaty backed up with efficient regional and national policies.

UNFCCC

The United Nations Framework Convention on Climate Change (UNFCCC) was created at the Rio Earth Summit in 1992 to negotiate a worldwide agreement for reducing GHGs and limiting the impact of climate change. The UNFCCC officially came into force on 21 March 1994. As of March 2014, UNFCCC has 196 parties. Enshrined within the UNFCCC are a number of principles

including agreement by consensus of all parties and differential responsibilities. The latter is because the UNFCCC acknowledges that different countries have emitted different amounts of GHGs and therefore need to make greater or lesser efforts to reduce their emissions. To represent this formally at the negotiations two different groups of parties have been recognized: Annex I countries, which include all the developed countries, and non-Annex I, which include the less developed and rapidly developing countries. Annex I was subsequently divided when some countries argued that their economies were in transition. As a result, the richest countries were placed in an additional category, Annex II. The UNFCCC pays heed to the principle of contraction and convergence. This is the idea that the largest emitters of GHGs contract the amount of pollution towards a designated per-capita emissions total. For example, at the moment in the USA each person emits ten times more carbon dioxide than a person in India. For global equality, the amount emitted per person should be the same. To obtain an atmospheric carbon dioxide level that is stable at 450 ppm countries would have to rapidly drop their per-capita emissions to 2 tonnes per person per year. Figure 31 shows how far away from that level many developed and developing countries are. In the developed world, to ensure that other countries can develop as rapidly as possible,

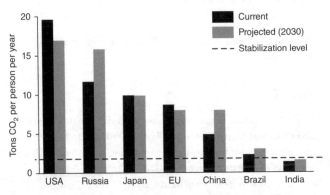

31. Historic and predicted per-capita carbon dioxide emissions

policy makers may have to consider an ultimate zero-carbon budget for their country. Some countries who can export renewable or alternative energy may even be able to produce a negative carbon economy. But considering the state of the climate change negotiations these discussions are not yet occurring.

Kyoto

Since the UNFCCC was set up, the nations of the world, 'the parties', have been meeting annually at the 'Conference of the Parties' (COPs) to move negotiations forward. Only five years after the UNFCCC was created at COP3 on the 13th December 1997, the first international agreement, the Kyoto Protocol, was drawn up. This stated the general principles for a worldwide treaty on cutting GHG emissions and, more specifically, that all developed nations would aim to cut their emissions by 5.2 per cent on their 1990 levels by 2008–12. The Kyoto Protocol was ratified and signed in Bonn on 23 July 2001, making it a legal treaty. However, the USA, under the leadership of President Bush, withdrew from the climate negotiations in March 2001 and so did not sign the Kyoto Protocol at the Bonn meeting. With the USA producing about one-quarter of the world's carbon dioxide pollution, this was a big blow for the treaty. Moreover, the targets set by the Kyoto Protocol were reduced during the Bonn meeting to make sure that Japan, Canada, and Australia would join (see Box 3). Australia finally made the Kyoto Protocol legally binding in December 2007.

The treaty did not include developing countries. This was to balance out the historic legacy of emissions by developed countries (Figure 4). It was then assumed that developing countries would join the post-2012 agreement. The Kyoto Protocol came into force on 16 February 2005, after Russia ratified the treaty, thereby meeting the requirement that at least 55 countries, representing more than 55 per cent of the global emissions, had signed up to it. Russia's membership tipped the scales, which allowed the Kyoto Protocol to become international

Box 3 UNFCCC negotiating alliances

Below is a 'who's who' guide to the international climate talks.
These different coalitions, formed during the climate change
negotiations, provide us with some insight into the differing
agendas of various countries. In addition, there are strong
lobbying interests from both individual states and environmental,
business, and industrial groups, which are also discussed below.

Annex I and Annex II (developed countries)

USA

USA is the key actor amongst developed countries. Thirty per cent
of all anthropogenic carbon dioxide was produced in the USA and
currently they have one of the highest per-capita emissions at 18
tonnes of carbon per year. Despite some of the best climate
science being produce in the USA, they have been only reluctantly
involved in the UNFCCC process. In fact they withdrew from the
Kyoto process and then at Copenhagen by-passed the democratic
process and instigated the non-binding Copenhagen Accord. The
negotiations have not progressed as far as expected over the last
20 years due to the reluctance of the USA to adopted binding
targets, hence China, India, and other rapidly developing
countries also see no reason why they should limit their
emissions.

JUSSCANNZ

This group of non-EU OECD (Organisation for Economic
Co-operation and Development) countries acted as a loose
information-sharing coalition during the UNFCCC negotiations,
lacking coordinated positions. JUSSCANNZ stands for Japan, USA,
Switzerland, Canada, Australia, Norway, and New Zealand.
Iceland and other OECD countries, such as Mexico, often
attended the group meetings. The over-arching concern of
JUSSCANNZ has always been the cost of tackling climate change.

The group is, however, split. Japan, New Zealand, Norway, and Iceland already enjoy high-energy efficiency and/or an energy mix dominated by low-carbon sources. The GHG emissions per unit of GDP and per capita are, therefore, much lower than the OECD average, so their main concern is the cost of abatement. The second group is Australia, Canada, and the USA—the so-called 'New World' countries—who face very different national circumstances with relatively low energy efficiency and an energy mix dominated by fossil fuels, growing populations, and large geographical areas, all of which lead to high emissions per unit of GDP and per capita. These countries' main concern is the cost of mitigating climate change in changing their energy-intensive infrastructure.

EU

The EU has maintained a coordinated position on climate change, usually speaking through its presidency, which rotates every six months. It has been rare for individual EU states to speak during the UNFCCC negotiations. The EU has a very similar split in its members to JUSSCANNZ, with both high and low energy-efficient economies. The consensus view of the EU has been to position itself as the environmental leader, with the mandated cuts of 20% by 2020, which at Copenhagen they were prepared to raise to 30% if other countries signed up to the post-2012 agreement. The EU rationale has been that any negotiated reduction could then be apportioned between the EU countries, depending on their development. This position has been greatly aided by both the UK and Germany experiencing a significant downturn in GHG emissions. In the UK, this was done by replacing coal with gas, while Germany's downturn was due to updating and cleaning up the inefficient industries of the former East Germany. However, the internal divisions within the EU and its cumbersome internal decision-making procedures make it a frustrating negotiating partner.

Non-Annex I (developing countries)

G77 and China

The Group of 77 (G77) is the main developing-country coalition and was formed in 1964 during the New International Economic Order negotiations under the UN Conference on Trade and Development (UNCTAD). China regularly allies itself with this group, which now has over 130 members. The Group operates according to a consensus rule. Without consensus, that is all countries within this group agreeing, no common position is articulated. Given the wide variety of interests that the G77 encompasses, however, it has been common for individual parties and groups also to speak during the UNFCCC negotiations, even when there was a common position. G77 symbolizes the North–South divide, with G77 seeing climate change as essentially an issue about development. Two major concerns are expressed by this group: first, that poor countries' development will be hindered by having to reduce emissions; and, second, that carbon trading must be allowed as a way of boosting income to developing countries.

China

In the first phase of the UNFCCC negotiations China was very clear that developed countries should make the first emissions reduction commitments given their historic legacy. With the huge expansion of the Chinese economy over the last 15 years China has now become a large emitter of GHGs. This is despite the fact that they have invested heavily in wind, hydro, and nuclear energy. China has also developed energy efficiency targets to lower energy use (emissions per unit of GDP) by 16% by 2015. China has now been pushing for binding targets from developed countries but has also provide its own conditional commitments.

AOSIS

The Alliance of Small Island States (AOSIS) was formed in 1990 during the Second World Climate Conference to represent the

interests of low-lying and small island countries that are particularly vulnerable to sea-level rise. It comprises some 43 states, most of which are also members of the G77. This group has regularly spoken at the UNFCCC negotiations, often but not always through its chair, although individual countries have also intervened. The AOSIS position has always been to achieve the tightest control on global emissions, as their countries seem to be most at threat from the impacts of global warming.

OPEC

The Organization of Petroleum Exporting Countries (OPEC) regularly, informally coordinate their positions in the climate change negotiations but have never spoken as a united group. The central position of this group is the protection of their main economic export, oil, and the prevention of any treaty that undermines the significant usage of fossil fuels.

African Group

The African Group is a formal regional group under the UN system, but it has only sometimes intervened during negotiations. More often, countries within this group have spoken for themselves or through the coordinating role of the G77. The African Group has been used mainly for ceremonial statements.

Non-state Actors

ENGOs

The Environmental Non-Governmental Organizations (ENGOs), though not a homogeneous group, they have a relatively united view on climate change. They universally accepted the science of climate change and its possible impact, and have campaigned for strong commitments on the part of governments and business to address the problem. However, there are significant differences among the ENGOs regarding specific issues in negotiations, particularly the possibility of emissions trading. The split can be

seen as reflecting a cultural difference between the New and Old Worlds. For example, Greenpeace International, based in Amsterdam, is strongly opposed to emissions trading, while Brazilian Friends of the Earth are strongly supportive of it.

BINGOs

Business and Industry Non-Governmental Organizations (BINGOs) were another powerful lobby at the UNFCCC negotiations. However, unlike the ENGOs, they are a diverse and loose-knit group, with three main sub-groups. At the more progressive end of the spectrum lie 'green' business, including the 'sunrise' renewable energy industries and insurance companies, who recognized climate change as a potential business opportunity and urged decisive action on the part of governments. The middle ground was occupied by the group which accepted the science of climate change but called for a prudent, cautious approach to mitigation. At the other extreme are the fossil-fuel lobby, mostly US-based industries such as the Global Climate Coalition. These were known as the grey BINGOs or the carbon club, who supported only the weakest action on climate change, stressing the economic costs and scientific uncertainties. Some of these BINGOs openly opposed the negotiations and have worked with OPEC states to block progress in both the IPCC and the climate change negotiations.

law. The Kyoto Protocol provided 38 industrialized nations with GHG emissions reduction targets. A total $500 million (£350 million) fund per year was to be provided by the industrialized world to help developing countries adapt to climate change and provide new clean technologies. It also set up the Clean Development Mechanism (CDM) so that developed countries could invest in and gain from a carbon credit in a developing country project. The lack of multi-level governance meant that

the CDM did not work as well as was expected. For example, credits for projects involving the capture of industrial gases (hydrofluorocarbons or HFCs) have been regrettably easy to game. The regulation has created a perverse incentive for companies to produce more HCFC-22, a refrigerant and powerful GHG being phased out under the Montreal Protocol, in return for windfall profits for capturing the HFC-23 by-product from its production. About 70 per cent of Certified Emission Reductions in the CDM have come from projects of this kind. Depressingly, the European Commission concluded in 2012 that production of HCFC-22 would have been lower today if the CDM had been absent.

Copenhagen

There were huge expectations of COP15 (Copenhagen) in 2009. New quantitative commitments were expected to ensure a post-2012 agreement to seamlessly move on from the Kyoto Protocol. Barack Obama had just become president of the USA. The EU had prepared an unconditional 20 per cent reduction of emissions by 2020 on a 1990 baseline and a conditional target rising to 30 per cent if other developed countries adopted binding targets. Most other developed countries had something to offer. Norway was willing to reduce emissions by 40 per cent and Japan by 25 per cent from a 1990 baseline. Even the USA offered a 17 per cent reduction on a 2005 baseline, which was an equivalent drop of 4 per cent on a 1990 baseline. But the Copenhagen conference went horrible wrong. First the Danish government had completely underestimated the interest in the conference and provided a venue that was too small. So in the second week, when all the high-powered country ministers and their support arrived, there was not enough room, meaning that many NGOs were denied access to the negotiations. Second, it was clear that the negotiators were not ready for the arrival of the ministers and that there was no agreement. This led to the leaking of the 'The Danish Text',

subtitled 'The Copenhagen Agreement', and the proposed measures to keep average global temperature rise to 2°C above pre-industrial levels. It started an argument between developed and developing nations as it was brand new text that had just appeared in the middle of the conference. Developing countries accused the developed countries of working behind closed doors and making an agreement that suited them without seeking consent from the developing nations. Lumumba Stanislaus Di-Aping, chairman of the G77, said, 'It's an incredibly imbalanced text intended to subvert, absolutely and completely, two years of negotiations. It does not recognize the proposals and the voice of developing countries.'

The final blow to getting an agreement on binding targets came from the USA. Barack Obama arrived only two days before the end of the conference, he convened a meeting of the USA with the BASIC (Brazil, South Africa, India, and China) countries, while seeking no discussions with the other UN nations, and he created the Copenhagen Accord. The Copenhagen Accord recognizes the scientific case for keeping temperature rises below 2°C, but it does not contain a baseline for this target, nor commitments for reduced emissions that would be necessary to achieve the target. Earlier proposals that would have aimed to limit temperature rises to 1.5°C and cut CO_2 emissions by 80 per cent by 2050 were dropped. The agreement made was non-binding and countries had until January 2010 to provide their own voluntary targets. It was also made clear that any country that signed up to the Copenhagen Accord was also stepping out of the Kyoto Protocol. Hence the USA was able to move away from the binding targets of Kyoto Protocol, which should have been enforced until 2012, and fostered a weak voluntary commitment approach. The Bolivian delegation summed up the way the Copenhagen Accord was reached—'anti-democratic, anti-transparent and unacceptable'. It is also not clear what the legal status of the Copenhagen Accord is as it was only 'noted' by the parties, not agreed, as only 122, subsequently rising to 139 countries, agreed to it.

The UNFCCC negotiations has taken another blow as trust has been eroded when, in January 2014, it was revealed that the US government negotiators had information during the conference obtained by eavesdropping on meetings of other conference delegations. Documents leaked by Edward Snowden showed how the US National Security Agency (NSA) had monitored communications between countries before and during the conference. The leaked documents show that the NSA provided US delegates with advance details of the Danish plan to 'rescue' the talks should they founder, and also about China's efforts before the conference to coordinate its position with that of India.

Post-Copenhagen

The failure of COP15 in Copenhagen and its voluntary commitments has cast a long shadow over the successive COP meetings. Including the revelation by Wikileaks that US aid funding to Bolivia and Ecuador was reduced because of their opposition to the Copenhagen Accord. At COP16 in Cancun and COP17 in Durban the UNFCCC negotiations were slowly put back on track with the aim to get legally binding targets. Also significant progress was made in the REDD+ (Reduced Emissions from Deforestation and Forest Degradation including safeguards for local people), which is discussed later in this chapter. It was, however, at COP18 in Doha in December 2012 that a second commitment period starting on 1st January 2013 was agreed, to last eight years. This ensured that all Kyoto mechanisms and accounting rules remained intact for this period and parties have reviewed their commitments with a view to increasing them, if possible, at the end of 2014. A guideline of reduction of between 25 and 40 per cent of the 1990 baseline by 2020 has been recommended. The second major outcome was the adoption of a timetable for a binding climate agreement, which is to be agreed at COP21 in Paris in 2015. This will cover both developed and developing countries, and will come into force by 2020. So the failure of COP15 in Copenhagen and the current lack of new

internationally binding emission targets should in the next few years be set straight again, barring any further political sabotage at the Paris COP in 2015.

Is the UNFCCC process flawed?

Not far enough. The first major flaw in the UNFCCC procedure is that it has failed to deliver any lasting agreement and it has been criticized for not going far enough with the suggested targets. This is based on the scientific view that a global cut of up to 60 per cent on a 1990 baseline is required to prevent major climatic change by 2050. If room is to be left for development, it would mean that the developed world would have to cut emissions by at least 80 per cent.

No enforcement. The fundamental problem with international agreements and treaties is there there are no real means of enforcement. This was one of the arguments that the USA used when proposing the Copenhagen Accord, suggesting that even binding targets must in effect be voluntary as countries decide whether or not to comply. This is why policies and laws are required at a regional level, such as the EU, and at a national level, such as the UK's Climate Change Act. The only way to translate international treaties is through regional and national laws. This multi-level governance is also required to stop gaming of particular systems.

Green colonialism. Many social and political scientists have raised philosophical and ethical doubts about climate negotiations as a whole. The main concern is that they reflect a version of colonialism, since rich developed countries are seen to be dictating to poorer countries how and when they should develop. Countries such as India and China have resisted calls to cut their emissions, stating that it would damage their development and attempts to alleviate poverty. Others have supported measures such the CDM as they provide a development dividend moving

money from the rich to the poorer countries. Again, however, since 80 per cent of the project credits are allocated to China, Brazil, India, and Korea, which are among the richest developing countries, funding is still not necessarily reaching the world's poorest. Also, 60 per cent of carbon credits have been purchased by the UK and the Netherlands, resulting in a very skewed financial exchange. The moral high ground of supposedly anti-green colonialism was employed by the EU and international NGOs during the Kyoto Protocol process to block the suggestion of global carbon trading. They felt that those who had polluted the most should be the first to cut back. However, national NGOs such as Greenpeace Brazil, some developing countries, and the USA argued strongly that global carbon trading was the only way forward to ensure everyone signed up to the Kyoto Protocol. In many ways the then US president, Bill Clinton, knew the only way to get emissions reduction policies through Congress was to include a trading mechanism. As we know, the moral high ground won and only CDMs were included. It led to the USA's withdrawal from the Kyoto Protocol. What is interesting is that the EU then realized it could not cut its own emissions the traditional way and thus set up the European Emissions Trading Scheme (ETS) for all facilities above 20 megawatts in electricity, ferrous metals, cement, refineries, pulp and paper, and glass industries, which together represent over 40 per cent of the EU total emissions.

Nation verses sector approach. The UNFCCC approach has another problem, which is embedded in the concept of the nation-state and is a major issue in a global capitalist world with supposedly free trade. For example, if the USA through the Copenhagen Accord wants to reduce carbon emissions from heavy industry, it could impose a carbon tax on steel and concrete production. However, if other countries in the world do not have this restriction, their products become cheaper, even including the cost of transportation by ship, air, or road to the USA, all of which would lead to the emission of more carbon dioxide overall. So

global economics can undermine any national attempts to do the right thing and reduce their emissions. An alternative approach would be for global agreements to be made at a sector level. For example, there could be a global agreement on how much carbon can be emitted per ton of steel or concrete produced. All countries could then agree only to buy steel or concrete produced in this low emission way, which would make for a fairer trading scheme, with countries not losing out as a result of changes within their industries to lower GHG emissions. The problems are, of course, how to police such a scheme across so many different industrial sectors.

A unilateral approach

With the realization that the UNFCCC negotiations have stalled, a number of countries and regions have proposed carbon cuts way above those currently negotiated. The EU Council of Ministers has agreed the 20:20:20 targets for 2020. That is, 20 per cent renewable energy for all 27 EU countries; a 20 per cent improvement in energy efficiency; and a 20 per cent cut in total GHGs. The first national climate change legislation mandating a targeted cut in GHG emission was enacted in 2008 in the UK with their Climate Change Act. This binds the UK government to reducing the country's carbon emissions by 80 per cent, using a 1990 baseline, by 2050. All three main political parties supported this Act. As part of this new set of legislations the Climate Change Committee was set up, independent of government, to monitor and advise successive governments on achieving this goal. Every five years they are to provide carbon budgets for the sitting government and advice on how to achieve these interim targets. This shows international independent leadership that if the world's fifth largest economy can do it, then it is possible for other countries to follow. In 2012 Mexico became the second country in the world and the first developing country to enshrine GHGs

reduction into national law. Mexico have mandated reducing GHG emissions by 30 per cent by 2020, and by 50 per cent by 2050. Other countries are also starting to develop climate and energy laws with respect to cutting GHG emission. The Global Legislators Organisation (GLOBE International) and the Grantham Research Institute at the London School of Economics published a study in 2014 that audited climate legislation across 66 countries, all of which produce 88 per cent of anthropogenic GHG emissions. The study found that 64 out of 66 countries had enacted or were planning to enact significant climate or energy legislation. In all, more than 500 climate laws are now on the books in the nations most responsible for climate change. Emerging economies like Mexico and China are leading the charge, and 'flagship legislation' was passed in eight countries, most of them developing nations such as Bolivia, El Salvador, and Mozambique. The study also shows that some countries such as Australia and Japan have 'backslid' and started to reverse climate legislation. But though the study brings good news, it also found that current national policies around the world still fall short of the goal of limiting a global temperature rise to no more than 2°C. This suggests that internationally binding targets are required to help focus and enhance these national laws.

Key regions have also produced climate change laws. California has mandated a 60 per cent reduction by 2050. This may be very significant, as there is a history of innovative environmental laws in California being accepted as Federal laws in due course. If the Californian model were adopted by the whole of the USA, this would make a huge difference to global emissions. Another interesting development in the USA is that carbon dioxide has now legally been defined as a pollutant, which means the Environmental Protection Agency (EPA) now have jurisdiction over its production, and could enact control measures without recourse to Federal government.

Carbon trading

Many politicians have advocated either a regional or a global carbon trading scheme. The most successful system of carbon trading is 'cap and trade', whereby politicians set a cap, a maximum total of pollution allowed, and a trading system is then set up so that different industries can trade credits. It is acknowledged that the various industries can clean up at varying rates and costs, and this trading system allows the most cost-effective approach to be found. This type of system has already been used in US emissions trading to reduce sulphur dioxide and nitrous oxides, the primary components of acid rain, and it has been highly successful. The Clean Air Act of 1990 required electrical utilities to lower their emissions of these pollutants by 8.5 million tonnes compared with 1980 levels. Initial estimates in 1989 suggested it would cost $7.4 billion; a report in 1998 based on actual compliance data suggested it had cost less than $1 billion.

The EU's ETS to reduce GHGs has been successful. The EU's ETS covers more than 11,000 installations with a net energy use of 20 megawatts and includes electricity generation, ferrous metal production, cement production, refineries, pulp, paper, and glass manufacturing. The ETS covers 31 countries, which include all 28 EU member states plus Iceland, Norway, and Liechtenstein. The ETS covers half the EU's carbon dioxide emissions and 40 per cent of its total GHG emissions. Under the 'cap and trade' principle, a cap is set on the total amount of GHGs that can be emitted by installations in each country. 'Allowances' for emissions are then auctioned off or allocated for free, and can subsequently be traded. Installations must monitor and report their carbon dioxide emissions, ensuring they hand in enough allowances to the authorities to cover their emissions. If emission exceeds what is permitted by its allowances, an installation must purchase allowances from others. Conversely, if an installation has performed well at reducing its emissions, it can sell its leftover

credits. This allows the system to find the most cost-effective ways of reducing emissions without significant government intervention. The scheme has been divided into a number of 'trading periods'. The first ETS trading period lasted three years, from January 2005 to December 2007. The second trading period ran from January 2008 until December 2012, coinciding with the first commitment period of the Kyoto Protocol. The third trading period began in January 2013 and will last until December 2020. Compared to 2005, when the EU's ETS was first implemented, the proposed caps for 2020 represents a 21 per cent reduction of GHGs.

Controversially, the ETS tried in 2012 to extend its carbon market to include international aviation emissions. It stipulated that only aircraft not included in another carbon trading schemes would be included. As there are currently no other schemes it would have included 60 per cent of the world's international commercial aviation emissions. The airline industry and countries including China, India, Russia, and USA reacted adversely to the inclusion of the aviation sector. They argued that the EU did not have jurisdiction to regulate flights when they were not in European skies; and China and the United States threatened to ban their national carriers from complying with the scheme. On 27 November 2012, the United States enacted the European Union Emissions Trading Scheme Prohibition Act, which prohibiting US carriers from participating in the European Union Emission Trading Scheme.

The EU's ETS works as it includes a group of developed countries and the emissions allowances are agreed in advance and take into account the different countries' energy mix and level of development. The ETS also limits how many carbon credits can be bought from other projects from developing countries through mechanisms such as the CDM. This is because there is a large cost differential in cutting carbon emissions between the developed and developing world: it is much cheaper to cut or avoid

emissions in developing countries. Hence a global carbon trading scheme would be highly problematic due to a number of key reasons. For one thing, it act as a form of colonialism, where rich countries could maintain their levels of consumption while getting credit for carbon savings in inefficient industrial projects in poorer countries. Companies in both China and India have already been found to be playing the system by significantly increasing their GHG emissions to gain carbon credits. Other criticisms of a possible global trading scheme include the fact that these schemes create new uncertainties and risks, which can be commodified by means of derivatives, thereby creating a new speculative market, which as we have seen have caused huge problems with food prices.

REDD+

The idea of developing an instrument on deforestation within the climate change negotiations was first suggested at COP11 (2005) in Montreal and was referred to as RED (Reduced Emissions from Deforestation). The UN REDD Programme (Reducing Emissions from Deforestation and Forest Degradation) was agreed in principle at COP13 (2007) in Bali. This has been subsequently refined as REDD+, the '+' representing safeguards to protect local people and REDD++, with the second '+' representing safeguards to the local ecosystem and biodiversity. REDD++ is visualized as a win-win solution that can protect forests and ecosystems, promote reforestation, and protect and compensate forest dwellers from/ for lost income from exploiting their forested land. But REDD++ has not been well designed and there is still discussion on a whether it is fund or market driven process. There have also been problems dealing with the issue of 'leakage'. Leakage refers to the fact that forest maintenance in one place may lead to deforestation in another. Hence REDD++ needs to be adopted at a national level so that domestic leakage can be avoided. There is also the problem of 'additionality' because pristine rainforest is already absorbing a lot of atmospheric carbon dioxide, part of the

25 per cent of our pollution take up is by the land biosphere. So there is no additional benefit from that forest except if it is under threat and then, if cut down, it would release all that stored carbon. So REDD++ projects would have to show additional carbon benefits. The final problem is how to find an economic way to measure, report, and verify (MRV) carbon stored in protected forest or gained in reforested areas. Currently a combination of satellite data with ground truthing seems the most cost effective approach.

Despite these problems REDD has built up considerable momentum and relevant policies are being developed in 40 countries to make them REDD-ready. Considerable funding has been made available for this through World Bank's Forest Carbon Partnership Facility, Norway's International Climate and Forest Initiative, the Global Environment Facility, Australia's International Forest Carbon Initiative and the Collaborative Partnership on Forests. Some successes have already occurred, for example the bilaterial agreement between Norway and Indonesia led to a two-year ban on government issued licences to convert forest-land to other uses. So the confusion in REDD++ design and lack of strong leadership through the UNFCCC process has not stopped countries pushing ahead with this approach.

Summary

Climate change can only be solved through binding, international agreements to cut GHG emissions. It is clear that because GHG emissions are rising so quickly developing as well as developed countries will have to agree emission targets. Figure 32 shows that even if developed countries completely cease all GHG emissions by 2040 the emission from developing countries would break through the 450 ppm limit. With hindsight the UNFCCC Kyoto Protocol must be recognized as a ground-breaking agreement, as it had binding targets and was signed by over 190 countries. The failure of COP15 at Copenhagen and thus the introduction by the

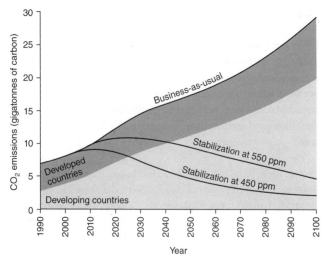

32. **Predicted carbon dioxide emissions for business-as-usual and stabilization at atmospheric concentrations of either 550 ppm or 450 ppm**

US and BASIC countries of voluntary pledges set back the international negotiations by at least ten years. COP18 in Doha brought the negotiations back on track, including reinstating the Kyoto mechanisms and accounting rules, and encouraging the parties to review and if possible increase their commitments. A timetable for a binding climate agreement was also agreed for finalization at COP21 in Paris in 2015, covering both developed and developing countries, and due to come into force by 2020. Again, as with COP15 in Copenhagen, expectations are running high for an agreement at COP21 in Paris. However, the big question is whether or not the USA will engage with it this time, or again disrupt the negotiations in order to delay any binding targets. History, unfortunately, leads us to be pessimistic, as the USA has failed to ratify the Rio pact on biodiversity, withdrawn from the anti-ballistic missile treaty, opposed the ban on landmines, opposed amendments to the biological warfare

convention, opposed the setting up of an international criminal court, and sidelined the UN in the lead-up to the second Iraq war. At COP13 (2007) in Bali the Papua New Guinea delegation famously told the US delegation: 'If you're not willing to lead, please get out of the way.' A short time later, the USA reversed its blocking position. We can but hope that international pressure is such that the USA will now finally decide that environmental security is essential and that climate change must be dealt with.

Chapter 8
Solutions

Introduction

At the moment there is no binding international agreement to cut emissions though many countries are unilaterally undertaking mitigation policies. It is clear that the current trend of GHG emissions is now above the worst-case scenarios considered by the 2013/14 IPCC reports. This is primarily due to the rapid development of the emerging economies such as China, India, Brazil, and South Africa. So this chapter examines three types of solutions to climate change. The first is adaptation, which simply put is providing protection for the population, as we already know that there will be climate change even if emissions were radically cut back to 1990 levels. Second is mitigation, which in its simplest terms is reducing our carbon footprint and thus reversing the trend of ever-increasing GHG emissions. Third is geoengineering that involves large-scale extraction of carbon dioxide from the atmosphere or modification of the global climate.

Adaptation

There will certainly be climate change. Many countries will be adversely affected in the near future, and nearly all countries will be affected in the next 30 years. So the second report of the IPCC

Fifth Assessment published in 2014 examines the impacts of climate change and the potential sensitivity, adaptability, and vulnerability of each national environment and socioeconomic systems. For example, as flooding is going to become more prevalent in Britain, damage to property and loss of life could be prevented with new flood defences and strict new laws that limit building on floodplains and vulnerable coasts.

The IPCC believes there are six reasons why we must adapt to climate change: (1) climate change cannot be avoided (see Chapter 4); (2) anticipatory and precautionary adaptation is more effective and less costly than forced last-minute emergency fixes; (3) climate change may be more rapid and more pronounced than current estimates suggest, and unexpected and extreme events are likely to occur; (4) immediate benefits can be gained from better adaptation to climate variability and extreme atmospheric events (for example, with the storm risks, strict building laws and better evacuation practices would need to be implemented); (5) immediate benefits can also be gained by removing maladaptive policies and practices (for example, building on floodplains and vulnerable coastlines); and (6) climate change brings opportunities as well as threats. Figure 33 provides an example of how countries can adapt to predicted sea-level rise.

The major threat from climate change is its unpredictability (see Chapter 6). Humanity can live in almost any extreme climate, from deserts to the Arctic, but we can only do so when we can predict what the extremes of the weather will be. So adaptation is really the key to dealing with climate change, but it must start now, as infrastructure changes can take up to 30 years to implement. For example, if you want to change land-use, for example, by building better sea defences or returning farmland back to natural wetlands in a particular area, it can take ten years to research and plan the appropriate measures. It can then take another ten years for the full consultative and legal processes, and

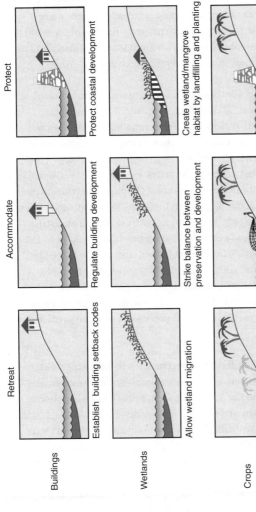

Climate Change

	Retreat	Accommodate	Protect
Buildings	Establish building setback codes	Regulate building development	Protect coastal development
Wetlands	Allow wetland migration	Strike balance between preservation and development	Create wetland/mangrove habitat by landfilling and planting
Crops	Relocate agricultural production	Switch to aquaculture	Protect agricultural land

33. Model response strategies for future sea-level rise

138

a further ten years to implement these changes. It can then take another decade for the natural restoration to take place to complete the adaptation project (see Figure 34). A good example of this is the Thames Barrier, which currently protects London from flooding. It was built in response to the severe flooding in 1953 but did not open officially until 1984, 31 years later. Since then the UK Environmental Agency has had a programme of continually upgraded all the flood defences along the River Thames to make the barrier as effective as possible.

There are, however, limits to adaptation. First, climate change in particular areas may be such that it goes beyond our ability or finances to protect the population living there. Second, in some regions our ability to predict the effects of climate change are limited and thus formulating an adaptation plan is difficult. In Chapter 4, the Mekong River Basin example was discussed and our current inability to predict whether climate change will increase or decrease its annual discharge. Advising policy makers becomes extremely hard when the uncertainties do not even allow one to tell if a river catchment system in the future will have more or less water. But this lack of knowledge should also be communicated to policy makers so they understand the whole range of possible local climate scenarios they may face.

The other problem is that adaptation requires money to be invested now; many countries just do not have the money, and where they do, people do not want to pay more taxes to protect themselves in the future; most people live for the present. This is, of course, despite the fact that all of the adaptations discussed will in the long term save money for the local area, the country, and the world. As a global society we still have a very short-term view, usually measured in a few years between successive governments. Hence the solutions to climate change will have to combine adaptations with the mitigation strategies discussed below. Central to the UNFCCC negotiations have been assisting technology transfer from developed to developing

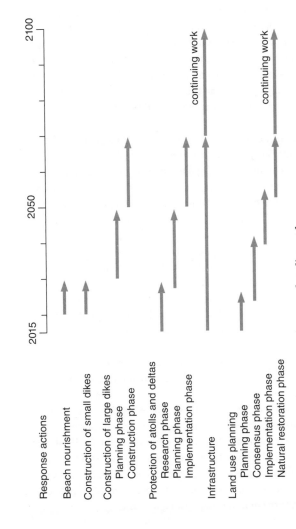

Climate Change

Response actions

Beach nourishment

Construction of small dikes

Construction of large dikes
Planning phase
Construction phase

Protection of atolls and deltas
Research phase
Planning phase
Implementation phase

Infrastructure

Land use planning
Planning phase
Consensus phase
Implementation phase
Natural restoration phase

2015   2050   2100

continuing work

continuing work

34. Lead times for response strategies to combat climate change

countries and setting up a climate change fund to help developing countries adapt.

The one thing that governments are doing now is setting up climate change impact assessments. For example, in the UK there is the UK Climate Impact Programme (<http://www.ukcip.org.uk>) that in January 2009 launched new products, based on the latest IPCC 2007 reports, showing the possible effects of climate change on the UK over the next 100 years. These products are aimed at the UK national and local government, industry, business, the media, and the general public. These will now be updated to the latest 2013/14 IPCC reports. If every government set up one of these programmes, then at least their citizens would have the opportunity to make informed choices about how their countries should be adapting to climate change.

Mitigation

The idea of cutting global carbon emissions by half in the next 30 years and by up to 80 per cent by the end of the century may sound extremely challenging; however, already the UK, Mexico, and California have made legally binding commitments to reduce carbon emissions by 80 per cent, 50 per cent, and 60 per cent, respectively, by 2050. Professors Steve Pacala and Robert Socolow, at Princeton University, published a very influential paper in the journal *Science* which makes this challenge seem more achievable. They took the business-as-usual emissions scenario and the desired 450 ppm scenario and described the difference between the two as a number of 'wedges'. So on this view, instead of seeing one huge insurmountable problem, really what we are faced with are between 16 and 20 medium-sized changes which add up to the desired big change (see Figure 35). They also provided several examples for the wedges, each of them approximately saving 1 GtC every year, as shown in Table 5. For example, one wedge would be doubling the efficiency of two billion cars from 30 miles per gallon (~9.4 litres per 100 km) to

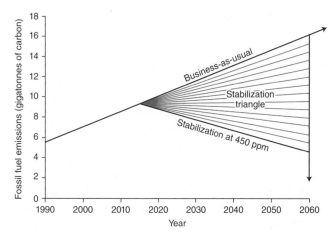

35. Stabilization wedges to achieve an atmospheric concentration of 450 ppm carbon dioxide

60 miles per gallon (~4.7 litres per 100 km), which is in fact a very achievable aim, as cars have already been built that can easily do 100 miles per gallon (2.8 litres per 100 km). Though Steve Pacala and Robert Socolow provide examples of what we can do to achieve the required cuts in GHGs, each country will have different strengths and weaknesses, and can apply the most suitable wedges for them. One of the first measures that all countries, industries, and individuals can take is to be more energy efficient. At the moment, the energy use in an average home in the USA is twice that of the average Californian home, while California's domestic energy use is twice that of Denmark. So already within the developed world there are huge savings to be made just by improving energy efficiency. It remains a puzzle why industry and business do not improve their energy use, as this can also significantly cut running costs. Efficiency gains, however, will ultimately be taken over by increased use. For example, if we did double the efficiency of two billion cars, this gain would be wiped out once another two billion cars are put onto the roads. So one of the most important areas is the production of clean, or carbon-free, energy, discussed later in this chapter.

Table 5. Princeton stabilization wedges

Option	Equivalent to 1 gigatonne of carbon per year wedge
Energy efficiency and conservation	
Economy-wide carbon intensity reduction (emissions/GDP)	Reduce carbon intensity by an additional 0.15% per year
1. Efficient vehicles	Increase fuel economy for 2 billion cars from 30 to 60 mpg
2. Reduced use of vehicles	Decrease car travel for 2 billion 30-mpg cars from 10,000 to 5,000 miles per year
3. Efficient buildings	Cut carbon emissions by one-fourth in buildings and appliances
4. Efficient coal plants	Produce twice today's coal power output at 60% instead of today's 32% efficiency
Fuel shift	
5. Gas power for coal power	Replace 1,400 gigawatts (GW) of coal plants with gas plants
CO_2 CCS	
6. Capture CO_2 at power plant	Introduce CCS at 800 GW coal or 1,600 GW natural gas plants
7. Capture CO_2 at coal-to-synfuels plants	Introduce CCS at synfuels plants producing 30 million barrels a day from coal
Nuclear fission	
8. Nuclear power for coal power	Add 700 GW (twice the current capacity)

(continued)

Table 5. Continued

Option	Equivalent to 1 gigatonne of carbon per year wedge
Renewable electricity and fuels	
9. Wind power for coal power	Add 2 million 1-megawatt-peak windmills (50 times the current capacity)
10. Photovoltaic (PV) power for coal power	Add 2,000 GW-peak PV (700 times the current capacity)
11. Biomass fuel for fossil fuel	Add 100 times the current Brazil or US ethanol production (one-sixth of world cropland)
Forests and agricultural soils	
12. Reduced deforestation, plus reforestation, afforestation, and new plantations	Decrease tropical deforestation to zero instead of 0.5 GtC emission per year, and establish 300 million hectares of new tree plantations (twice the current rate)
13. Conservation tillage	Apply to all cropland (ten times the current usage)

Alternative, renewable, or clean energy

Fossil fuels are an amazing discovery, and they have allowed the world to develop at a faster rate than it has at any other time in history. The high standard of living in the developed world is based on cheap and relatively safe fossil fuels. But as we have discovered burning fossil fuels has the unintended consequence of changing global climate. So in the 21st century we need to switch from fossil fuel energy to carbon-neutral energy. These include solar, wind, hydro, wave, and tidal energy. There are other reasons to switch to renewable sources of energy. First is the concern that we have reach 'peak oil' and 'peak coal' resource levels, and that the world is now running out of both. This is

highly unlikely given all the new reserves of oil and high-grade coal that have been discovered over the last 20 years. Nonetheless, as great colleagues of mine who are chemists keep on telling me, the greatest crime you can do with oil is to burn it, with our whole industrial society being built on plastics made from petroleum. Moreover, given the fracking revolution, there is now almost unlimited supply of natural gas. Second, countries have in the 21st century become very aware of 'energy security'; most developed countries' economies are heavily reliant on the import of fossil fuels, making them very vulnerable to volatility of the markets and international blackmail.

There now follows a brief discussion of the main alternative energy sources. Straight away we can see how different portfolios fit different countries. For example, the UK has the best wind resources in the whole of Europe, while Saudi Arabia has excellent conditions for solar power. Almost all the technology for these alternatives exist, except for nuclear fusion.

Solar. The Earth receives on average 343 W/m^2 from the Sun, and yet the Earth as a whole only receives a two-billionth of all the energy put out by the Sun. So the Sun is in many ways the ultimate source of energy, energy which plants have been utilizing for billions of year. At the moment we can convert solar energy directly to heat or electricity or we can capture the energy through photosynthesis by growing biofuels. The simplest approach is through solar heating. On a small scale, houses and other buildings in sunny countries can have solar heating panels on the roof, which heats up water, so people can have carbon-free hot showers and baths. On a large scale, parabolic mirrors are used to focus the solar energy to generate hot liquid (water or oil) to drive turbines to create electricity. The best places to situate solar heat plants are in low-latitude deserts, which have very few cloudy days per year. Solar heat plants have been built in California since the 1980s and are now being built in many other countries. Solar photovoltaic panels convert sunlight

directly into electricity. The individual rays of the Sun hit the solar panel and dislodge electrons inside it, creating an electrical current. The main advantage of solar panels is that you can place them where the energy is needed and avoid the complicated infrastructure normally required to move electricity around. Over the last decade there has been a massive increase in their efficiency, the best commercially available solar panels being about 21 per cent efficient, which is significantly more thanwith photosynthesis at about 1 per cent. Furthermore, there has been a significant drop in price due to huge investment in technology.

Biofuels. These are the product of solar energy converted into plant biomass via photosynthesis, which can then be used to produce a liquid fuel. The global economy is based on the use of liquid fossil fuels, particularly in the transport sector. So in the short term fuels derived from plants could be an intermediate low carbon way of powering cars, ships, and airplanes. There are concerns that production of biofuels could compete with that of food crops. In fact the recent food price peaks of 2008 and 2011/12 were initially blamed on biofuel production, but analysis of these massive increases showed that the real culprits were in fact increased oil prices and food speculation on the financial markets. Ultimately, electric cars are the future, because the required electricity can be produced carbon-neutrally. However, this energy source is not an option for airplanes. Traditional air fuel 'kerosene' combines relatively light weight with a high energy output. Research is being carried out to see whether a biofuel can be produced that is light enough and powerful enough to replace kerosene.

Wind. Wind turbines are an efficient means of generating electricity, if they are large and preferably located out at sea. Ideally, we need turbines the size of the Statue of Liberty for maximum effectiveness. The London Array is being built in the River Thames estuary and will generate 1,000 megawatts (MW), making it the world's largest consented offshore wind farm. When finished it could power up to 750,000 homes in Greater London

and reduce harmful carbon dioxide emissions by 1.4 million tonnes a year. However, the problem with wind turbines is twofold. First, they do not supply a constant source of electricity; if the wind does not blow, then there is no electricity. Second, people do not like them, finding them ugly, noisy, and a worry in terms of the effects they may have on local natural habitats. All these problems are easy to overcome by situating wind farms in remote locations, out at sea, and away from areas of special scientific or natural interest. Moreover, recent research has shown little or no effect on local wildlife even when close to land wind turbines. One study suggests that wind in principle could generate over 125,000 terrawatt-hours, which is five times the current global electricity requirement.

Wave and tidal. Wave and tidal power could also be an important source of energy in the future. The concept is simple: to convert the continuous movement of the ocean in the form of waves into electricity. However, this is easier said than done, and experts in the field suggest that wave power technology is now where solar panel technology was about 20 years ago—lots of catching up still required. But tidal power in particular has one key advantage over solar and wind—it is constant. In any country, for energy supply to be maintained at a constant level, there has to be at least 20 per cent production guaranteed, known as the baseline requirement. With the switch to alternative energy, new sources of this consistent baseline need to be developed.

Hydroenergy. Hydroelectric power is globally an important source of energy. In 2010, it supplied 5 per cent of the world's energy. The majority of the electricity comes from large dam projects. These projects can present major ethical problems as large areas of land must be flooded above the dam, causing mass relocation of people and destruction of the local environment. A dam also slows water flowing down a river and prevents nutrient-rich silt from being deposited lower down. If the river crosses national boundaries, there are potential issues over the rights to water and silt. For

example, one of the reasons why Bangladesh is sinking is the lack of silts due to dams on the major rivers in India. There is also a debate about how much GHG hydroelectric plants save, because even though the production of electricity does not cause any carbon emissions, the rotting vegetation in the area flooded behind the dam does give off significant amounts of methane.

Geothermal. Below our feet, deep within the Earth, is hot molten rock. In some locations, for example in Iceland and Kenya, this hot rock comes very close to the Earth's surface and can be used to heat water to make steam. This is an excellent carbon-free source of energy, because the part of the electricity you generate from the steam you use to pump the water down to the hot rocks. Unfortunately, it is limited by geography. There is, however, another way the warmth of the Earth can be used. All new buildings can have a borehole below them with ground-sourced heat pumps. Cold water is then pumped down into these boreholes and the ground warms the water up, cutting the cost of providing hot water to buildings and can be used almost everywhere in the world.

Nuclear fission. Energy is generated when you split heavy atoms such as uranium and this is nuclear fission. It has a very low direct carbon signature, but a significant amount of carbon is generated mining the uranium, building the nuclear power station, and decommissioning the power station and safely deposing of the nuclear waste. At the moment, 5 per cent of global energy is generated by nuclear power. The new generation of nuclear power stations are extremely efficient, producing nearly 90 per cent of the theoretically possible energy production. The main disadvantages of nuclear power are the generation of high-level radioactive waste and concerns about safety, though improvements in efficiency reduce waste and the new generations of nuclear reactors have state-of-the-art safety precautions built in. However, as the Chernobyl disaster in 1986 and the Fukushima Daiichi nuclear disaster in 2011 show, nuclear plants are still not

safe, being vulnerable to human error and natural disasters. The advantages of nuclear power, however, are that it is reliable and can provide the required base load in the energy mix, and it is technology that is already available and already thoroughly tested.

Nuclear fusion. This is the generation of energy that is found when two smaller atoms are fused together. It occurs in our Sun and every other star. The idea is that the heavy form of hydrogen found in sea water can be combined and the only waste product is the non-radioactive gas, helium. The problem, of course, is persuading those two atoms to join together. The Sun does it by subjecting the atoms to incredibly high temperatures and pressures. Some advances have been made at the Joint European Torus (JET) project in the UK, which has produced 16 megawatts of fusion power. The problem is the amount of energy required to generate the huge temperatures in the first place and the difficulty of scaling it up to a power plant size.

Carbon capture and storage

Removal of carbon dioxide during industrial processes can be tricky and costly, because not only does the gas need to be removed, but it must be stored somewhere as well. In the UK the world's first carbon capture and storage (CCS) demonstration project called White Rose has been commissioned. White Rose will be part of the Drax Power Station site near Selby, North Yorkshire, generating electricity for the UK as well as capturing approximately 2 million tonnes of carbon dioxide per year, some 90 per cent of all carbon dioxide emissions produced by the plant. The carbon dioxide will be transported through the National Grid's proposed pipeline for permanent undersea storage in the North Sea. The project is being financially supported by both the European Commission (~£250 million) and the UK government (~£100 million). This large-scale project is being carried out to demonstrate that the technology does work and that CCS has a key role to play in reducing future carbon dioxide emissions.

UK and EU financial support is required, as removal and storage costs could be somewhere between $10 and $50 per tonne of carbon dioxide. This would cause a 15 per cent to 100 per cent increase in power production costs. However, recovered carbon dioxide does not all need to be stored; some may be utilized in enhanced oil recovery, the food industry, chemical manufacturing (producing soda ash, urea, and methanol), and the metal-processing industries. Carbon dioxide can also be applied to the production of construction material, solvents, cleaning compounds, and packaging, and in waste-water treatment. However, in reality, most of the carbon dioxide captured from industrial processes would have to be stored. It has been estimated that theoretically two-thirds of the carbon dioxide formed from the combustion of the world's total oil and gas reserves could be stored in the corresponding reservoirs. Other estimates indicate storage of 90–400 Gt in natural gas fields alone and another 90 Gt in aquifers.

Oceans could also be used to dispose of the carbon dioxide. Suggestions have included storage by hydrate dumping—if you mix carbon dioxide and water at high pressure and low temperatures, it creates a solid, or hydrate, which is heavier than the surrounding water and thus drops to the bottom. Another more recent suggestion is to inject the carbon dioxide half a mile deep into shattered volcanic rocks in between giant lava flows. The carbon dioxide will react with the water percolating through the rocks. The acidified water will dissolve metals in the rocks, mainly calcium and aluminium. Once it forms calcium bicarbonate with the calcium, it can no longer bubble out and escape. Though if it does escape into the ocean, then bicarbonate is relatively harmless. With ocean storage there is the added complication that the oceans circulate, so whatever carbon dioxide is dumped, some of it will eventually return. Moreover, scientists are very uncertain about the environmental effects on the ocean ecosystems.

The major problem with all of these methods of storage is safety. Carbon dioxide is a very dangerous gas because it is heavier than

air and can cause suffocation. An important example of this was in 1986, when a tremendous explosion of carbon dioxide from Lake Nyos, in the west of Cameroon, killed more than 1,700 people and livestock up to 25 km away. Though similar disasters had previously occurred, never had so many people and animals been asphyxiated on such a scale in a single brief event. What scientists now believe happened was that dissolved carbon dioxide from the nearby volcano seeped from springs beneath the lake and was trapped in deep water by the weight of water above. In 1986, there was an avalanche that mixed up the lake waters, resulting in an explosive overturn of the whole lake, and all the trapped carbon dioxide was released in one go. However, at this very moment huge amounts of mined ancient carbon dioxide is pumped around the USA to enhance oil recovery. There are no reports of any major incidents and engineers working on these pipelines feel they are much safer than on the gas and oil pipelines, which run across most major cities.

Transport

One of the greatest challenges to mitigating GHG emissions is transport. At the moment, transport accounts for 13 per cent of GHG emissions globally. In the UK, the carbon emissions from energy production, business, and residential sectors are all going down despite annual growth in the economy; but even the UK government admits that transport emissions, mainly from cars, are growing at a formidable rate. Car growth over the next 20 years could wipe out all the cuts in carbon emissions made by the UK since 1990. If we extrapolate this to the rest of the world, we have everyone in the developing world aspiring to have the same standard of living as the West, and that includes at least one car per household and regular holidays by aeroplane.

In respect to cars, there are two possible solutions—biofuels and electricity. Biofuels have been discussed earlier, and mean that the current infrastructure of providing liquid fuel to cars could be

maintained. But as we have seen, biofuels must be carefully produced as they can compete for land-use with food production, can result in tropical deforestation, and can still be net emitters of carbon due to transport and production costs. Ultimately, electric cars are the future, because it can be guaranteed that the electricity produced is carbon-neutral. At the moment, we already have hybrid cars, which combine a petrol engine with a battery system. Porsche and McLaren have already produced the world's first hybrid supercars as the electric engine can generate much more torque producing 0 to 60 miles per hour in 2.5 and 2.8 seconds, respectively. For normal cars a hybrid system can improve engine efficiency and cut carbon emissions by an average of 50 per cent. It would be an important step forward if all new cars produced had to have this type of system. The next step would be to move to completely electric cars. This would require continual improvement in battery life and the building of infrastructure to allow cars to charge up—just as you charge up your mobile phone at home. In the long-term vision of the UK Climate Change Committee, the first step is to completely decarbonize electricity generation in the UK and then to increase production significantly to ensure the ground transport systems, cars, buses, and trains, are then 100 per cent electric (see Figure 36).

Aeroplanes have become an easy target for climate change campaigners as international flights have never been covered by an international treaty. The EU through the Emissions Trading Scheme did try to include emissions from commercial aviation emissions but this is now being fought in the international courts. At the moment, just 1.6 per cent of global emissions come from aviation. Research is being carried out to see whether a biofuel can be produced that is light enough and powerful enough to replace the traditional air fuel kerosene, though this seems a long way off at the time of writing. Hydrogen is not a solution, because its by-product is water; this is fine on the ground, but high up in the air it produces cirrus clouds which contribute to warming the

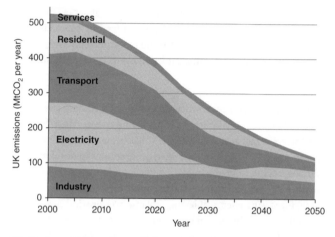

36. Envisaged UK carbon emissions to 2050

planet. Since at present there is no real fuel solution for aviation, the airlines are keen to be involved in carbon trading. This way, the airlines can 'off-set' their carbon emissions by ensuring an equivalent amount is saved elsewhere.

Carbon trading and offsetting

As discussed in Chapter 7, one of the most important tools in Europe to ensure that carbon emissions are lowered is carbon trading. This is one of two approaches that can be used to make alternative energy and CCS to be economically viable. The second approach is to penalize energy generation by fossil fuels with a trading scheme in which carbon permits must be bought. Since it is never cost effective to fit carbon capture to coal-fired power stations, this cost could be reduced by allowing these companies to trade the carbon saved. The European ETS and EU 20:20:20 laws are already making renewable sources of energy more competitive. So we must remember that if we really want to switch the global economy away from carbon and on to alternative energy sources and CCS, we need a fiscal method to

drive the markets. So far, the only approach that seems to have worked is within countries, and the use of trading blocks is the cap and trade system, allowing for the fact that reduction of carbon emission is very different in different countries and regions.

One of the most controversial aspects of carbon trading is offsetting, which has occurred through two systems: the UN CDM and the voluntary markets. The CDM has been described in Chapter 7 and involves UN-certified programmes in developing countries being funded to make significant GHG savings. These can include preventing methane release from an abandoned mine, increasing energy efficiency, solar power or wind power generation, or reduction in the manufacture of HFCs and CFCs. The CDM credits can be bought by countries, companies, or through the voluntary market. For example, every time you fly you can buy through a number of companies carbon offsets equivalent to the carbon emitted on your flight. In the West, a new branding of 'carbon-neutral' products has been seen. This seems to encompass anything from television companies such as BSkyB to paper manufacturers such as the Arjowiggins. Offsetting is controversial as it is argued that many of these cuts would have been made anyway, and also it means that companies and people may not be motivated to change their actual behaviours. On a practical level, however, it does offer a way that individuals, companies, and countries can make a difference. It also provides a means of dealing with unavoidable carbon emissions such as from aviation. What does need to happen is for there to be clear global rules on what is and is not an acceptable carbon offset. There also needs to be a clear verification service to ensure that the carbon saved is really saved.

Subsidies

One of the major political problems with reducing carbon emissions concerns energy subsidies. First, there are huge fossil

fuel subsidies, which continue to make oil, gas, and coal relatively cheap. Second, there is resistance to providing subsidies and tax incentives to the energy companies to build and supply renewable energy at competitive rates. This is mainly due to the current neoliberal view that states should not interfere with markets. However, this view concerning renewable energy subsidies usually comes from ignorance of the huge amounts of fossil fuel subsidies. In 2011, fossil fuel subsidies reached $90 billion in OECD countries and over $500 billion globally. This is compared to renewable energy subsidies, which reached only $88 billion globally. If we just focus on the USA, a study by the consulting firm Management Information Services estimated that between 1950 and 2010 the US government had given $369 billion to oil companies, $121 billion to natural gas companies, and $104 billion to coal companies. Oil also benefitted heavily from regulatory subsidies such as exemptions from price controls. Over the same period non-hydro renewable energy (primarily wind and solar) benefited from $74 billion in subsidies, largely in the form of tax policy and direct government expenditures on research and development (R&D). Nuclear power benefitted from $73 billion in subsidies, largely in the form of R&D, and hydro power received $90 billion in subsidies. According to the International Energy Agency, energy subsidies artificially lower the price of energy paid by consumers, raise the price received by producers, or lower the cost of production. 'Fossil fuels subsidies costs generally outweigh the benefits. Subsidies to renewables and low-carbon energy technologies can bring long-term economic and environmental benefits', according to Fatih Birol, Chief Economist at the International Energy Agency, without a phasing out of fossil fuel subsidies, we will have no chance of reaching any climate targets.

There is, however, another reason for the continuation of fossil fuel subsidies, and this is down to the ownership of the major oil and gas companies. Out of the top 26 oil and gas companies only 7 are private companies, the other 19 are fully or partly owned

by countries. Hence the state-owned companies are making huge amounts of money for the country and will continue to be given state aid in the form of subsidies and tax breaks to ensure that they are competitive with other oil and gas producing nations. This is only set to get worse with fracking and the shale gas revolution, with many countries such as the USA and UK having found new reserves of natural gas underground.

The top 26 oil and gas companies in 2014 in order of size are as follows (stars note a fully privately owned company): Saudi Aramco, Gazprom, National Iranian Oil, ExxonMobil*, PetroChina, BP*, Royal Dutch Shell*, Pemex, Chevron*, Kuwait Petroleum Corp, Abu Dhabi National Oil Co, Sonafrach, Total*, Petrobras, Rosneft, Iraqi Oil Minitery, Qatar Petroleum, Lukoil*, Eni, Statoil, ConocoPhillips*, Petroleos de Venezuela, Sinopec, Nigerian National Petroleum, and Petronas.

Geoengineering or technofixes

Geoengineering is the general term used for technologies that could be used to either remove GHGs from the atmosphere or to change the climate of the Earth (see Figure 37). Ideas considered under geoengineering range from the very sensible to the completely mad. At the moment we currently release over 8.5 GtC per year, so any interventions must operate on a very large scale.

Carbon dioxide removal. There are three main approaches to the removal and disposal of atmospheric carbon dioxide: biological, physical, and chemical.

1. Biological approaches on land include the use of biofuels, which were discussed above, and reforestation. Reforestation or afforestation and avoidance of deforestation are sensible win-win solutions. By maintaining our forest we can retain biodiversity, stabilize soils and local rainfall, and provide livelihoods for local people via carbon credits. An excellent example of this in action is

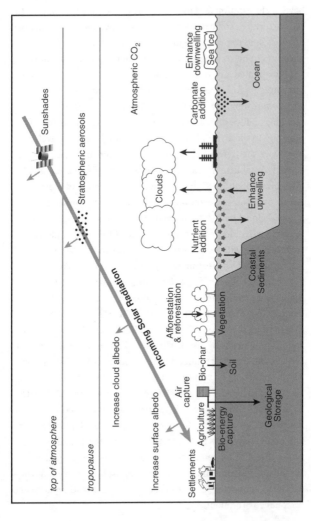

37. The range of geoengineering approaches

in China. By 1990 the Loess Plateau, the bread-basket for the China for at least the last 3,000 years, was turning into a dust bowl. Deforestation and over working of the soils had started to reduce fertility, so farmers were cutting down more trees to open up more land to produce enough food to survive. The Chinese government became aware of this problem and over the next ten years instigated a radical tree-planting programme with severe punishments for anyone caught chopping down trees. The effects were amazing: the trees stabilized the soils, greatly reducing soil erosion. The trees through transpiration added moisture to the atmosphere, reducing evaporation and water loss. Once the trees reached a critical size and area they also started to stabilize the rainfall. The land biosphere is already absorbing about 2 GtC per year of our pollution, and Steve Pacala and Robert Socolow estimate that if we completely stopped global deforestation and doubled our current rate of planting we could produce another one of their 1 GtC per year wedges along with all the win-win benefits that go with reforestation. Other estimates by Dr Yude Pan (US Forestry Service) and colleagues in *Science* suggest this could be much higher if we completely ceased deforestation and other land use changes (see Figure 38). In the UK, the Forestry Commission proposed an increase in the forested land in the UK from 12 per cent to 16 per cent by 2050. This would mean the government's target of 80 per cent reduction in carbon dioxide emissions by 2050 would only be 70 per cent due to the absorption and storage of carbon by our forests. The one area of the UNFCCC negotiations which are moving forward are the REDD+ discussion (see Chapter 7) whereby tropical developing countries will obtain payments for protecting existing forest which is under threat or for reforesting areas. REDD+ could make a significant contribution to reducing atmospheric carbon dioxide, as well as supporting ecosystem services and providing local and regional environmental security.

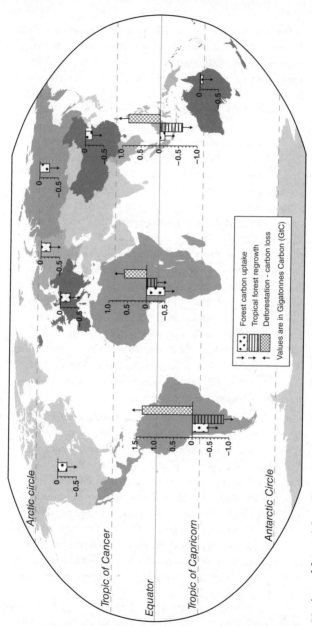

Solutions

38. Annual forest sinks and sources of carbon dioxide

2. The second biological approach is changing the uptake of carbon dioxide by the oceans. The most famous ocean 'technofix' was suggested by the late Professor John Martin. He suggested that many of the world's oceans are underproducing. This is because of the lack of vital micro-nutrients, the most important of which is iron, which allows plants to grow in the surface waters. Marine plants need minute quantities of iron, without which they cannot grow. In most oceans enough iron-rich dust gets blown in from the land, but it seems that large areas of the Pacific and Southern Oceans do not receive much dust and thus are barren of iron. So it has been suggested that we could fertilize the ocean with iron to stimulate marine productivity. The extra photosynthesis would convert more surface-water carbon dioxide into organic matter. When the organisms die, the organic matter drops to the bottom of the ocean, taking with it and storing the extra carbon. The reduced surface-water carbon dioxide is then replenished by carbon dioxide from the atmosphere. So, in short, fertilizing the world's oceans could help to remove atmospheric carbon dioxide and store it in deep-sea sediments. Experiments on this at sea have been highly variable, with some showing no effects at all while others have shown that the amount of iron required is huge. Also, as soon as you stop adding the extra iron, most of this stored carbon dioxide is released, as very little organic matter is allowed to escape out of the photic zone per year.

3. Physical. It is possible to remove carbon dioxide directly from the air. However, considering carbon dioxide makes up just 0.04 per cent of the atmosphere this is much harder than it sounds. One mad idea is the production of artificial or plastic trees. Klaus Lackner a theoretical physicist and Allen Wright an engineer, supported by Wally Broecker a climatologist, have designed carbon dioxide binding plastic, which can scrub carbon dioxide out of the atmosphere. The carbon dioxide is then released from the plastic and taken away for storage. The first problem is water, as the plastic releases the carbon dioxide into solution when wet, so the plastic trees would have to be placed in very arid areas or

require giant umbrellas. The second problem is the amount of energy required to build, operate, and then store the carbon dioxide. The third problem is one of scale; tens of millions of these giant artificial trees would be required just to deal with US carbon emissions.Why not just plant normal trees? However, it is clear that technology to remove carbon dioxide at source or ultimately from the atmosphere may be required.

4. Weathering. Carbon dioxide is naturally removed from the atmosphere over hundreds and thousands of years through the process of weathering, at a rate of 0.1 GtC per year, but this is 100 times less than what we are emitting. Only weathering of silicate minerals makes a difference to atmospheric carbon dioxide levels, as weathering of carbonate rocks by carbonic acid returns carbon dioxide to the atmosphere. By-products of hydrolysis reactions affecting silicate minerals are biocarbonates (HCO_{3-}), which are metabolized by marine plankton and converted to calcium carbonate. The calcite skeletal remains of the marine biota are ultimately deposited as deep-sea sediments and hence lost from the global biogeochemical carbon cycle for the duration of the lifecycle of the oceanic crust on which they were deposited.

There are a number of geoengineering ideas aimed at enhancing these natural weathering reactions. One suggestion is to add silicate minerals to soils that are used for agriculture. This would remove atmospheric carbon dioxide and fix it as carbonate minerals and biocarbonate in solution. However the scale this would have to be done is very large and there are unknown effects on soils and their fertility. Another suggestion is to enhance the rate of react of carbon dioxide with basalts and olivine rocks in the Earth's crust. Concentrated carbon dioxide would be injected into the ground and would create carbonates deep underground. Again, like many geoenigineering ideas, it is a great suggestion but very little work has been done to see if it is feasible, safe, and scalable.

5. Solar radiation management. As you can see from earlier discussions in the chapter, many of the ideas proposed as

geoengineering solutions are still just ideas and need a lot more work to see if they are feasible. This is particularly true of the solar radiation management ideas, many of which sound like something out of a really bad Hollywood B-movie. These suggestions include changing the albedo of the Earth, i.e. increasing the amount of solar energy reflected back into space to balancing the heating from global warming (Figure 37). These ideas also include erecting massive mirrors in space, injecting aerosols into the atmosphere, making crops more reflective, painting all roofs white, increasing white cloud cover, and covering large areas of the world's deserts with reflective polyethylene-aluminium sheets. The fundamental problem with all of these approaches is that we have no idea what effects they would have. At the moment, we are performing one of the largest geoengineering experiments ever undertaken by injecting huge amounts of GHG into the atmosphere, and though we know what in general will happen, we have no idea what the specific effects on our climate system could be. This is equally true of these geoengineering solutions—we currently have little idea if they would work or what unaccounted-for side effects they might have. In many ways, climate change for the Earth can be seen in the same way as illness and the human body: it is always preferable to prevent an illness than to try and cure one, and we all know the potential side effects of drug, chemo-, or radiation therapy.

Let us look at just one of these more far-fetched ideas as an example of what is wrong with solar radiation management: mirrors in space to deflect the sunlight. The most sophisticated of these suggestions is from Roger Angel, Director of the Centre for Astronomical Adaptive Optics at the University of Arizona, who suggests a mesh of tiny reflectors to bend some of the light away from the Earth. He himself admits this would be expensive, requiring 16 trillion gossamer-light spacecraft costing at least $1 trillion and taking 30 years to launch. Like all of the geoengineering ideas to change the Earth's albedo, it will not work. All these approaches are fixated with getting the

Earth's average temperature down and they miss the importance of the distribution of temperature with latitude, which is in fact what drives climate. In fact, Dan Lunt and colleagues at Bristol University have shown, using climate models, that these approaches take us to a completely different global climate, with the tropics being 1.5°C colder, the high latitudes 1.5°C warmer, and precipitation dropping by 5 per cent globally compared with pre-industrial times.

Geoengineering governance

The Royal Society 2009 report on geoengineering not only reviewed the current science but also made the important step of trying to understand the governance issues associated with playing with the global climate system. There are a great many ethical issues that arise when considering how changing regional and global climate may affect countries differently. There may be overall positive results but minor changes in rainfall patterns, which could mean that whole countries receive too little or too much rain, possibly resulting in disaster. The Royal Society summarized the current position, showing that there are three main views on geoengineering: (1) It is a route to buying back some time to allow the failed UNFCCC negotiations to catch up; (2) It represents a dangerous manipulation of the Earth system and may be intrinsically unethical; or (3) It is strictly an insurance policy against major mitigation policy failure. Even if research is allowed to go ahead and geoengineering solutions are required, like many emerging areas of modern technology, new flexible governance and regulatory frameworks will be required. Currently there are many international treaties with a bearing on geoengineering and it seems that no single instrument applies. Hence geoengineering like climate change challenges our nation-state view of the world and new ways of governing will be required in the future.

Summary

If we are to solve problems of climate change, we need to tackle two fundamental problems. The first is how we can reduce the amount of GHG pollution that we are emitting. At the moment we have no binding international treaty as many countries are reluctant to curb their emissions in case this damages their economy. This is because many nations aspire to have the same lifestyle and thus carbon footprint of Western countries. The world's population is currently just over seven billion and it is likely to rise and plateau at nine billion by 2050. That adds up to eight billion people who would like same lifestyle as a person living in the developed world, which would mean a huge potential increase in GHG emissions in this century to fuel this consumer dream. The second question concerns whether as a society we are prepared to invest the relatively small amount, about 1–3 per cent of world's GDP, according to Stern (2007), to offset a much larger bill in the future. If so, then we have the technology at the moment both to protect our population from climate change and to mitigate the huge predicted emissions of GHG over the next 100 years. As we have seen, energy efficiency, alternative energy sources, carbon trading, and offsetting all have a role to play. We must also consider 'disruptive technologies', that is, new technologies that we may not yet have even thought of that could change the way we produce or use energy. For example, most of us cannot think of life without a mobile phone or a computer, but this technology has been around for only a few decades; we can quickly become accustomed to change. There are also huge amounts of money to be made from opportunities surrounding changes to our energy use and our personal lifestyles, and as we will see in the next chapter, there may be many win-win situations whereby quality of life can be improved at the same time as stabilizing the climate of our planet.

Chapter 9
Envisioning the future

Introduction

Up to now this has been a standard book about climate change, reviewing the history, science, politics, and potential solutions. This chapter will, however, delve into some of the deeper questions of why humanity seems unable to deal with, on the face of it, a simple question of pollution. The challenge of climate change must be seen within the current dominant political and economic landscape. Only by understanding the fundamental societal and economic causes of carbon emissions can we hope to be able to build systems that can rapidly reduce them. At the same time as we deal with climate change we need to ensure that we also tackle other global challenges, such as poverty alleviation, environmental degradation, and global security. Future policies and international agreements need to provide win-win solutions that deal with the biggest challenges facing humanity in the 21st century.

Planetary limits

Climate change is but one of many global environmental problems that we currently face. One way to understand the current state of the global environment is by using the idea of planetary boundaries. This concept was proposed by a group of scientists led by Professor Johan Rockström from the Stockholm Resilience

Centre and Professor Will Steffen from the Australian National University. In 2009, the group proposed a framework of 'planetary boundaries' designed to define a 'safe operating space for humanity' for the international community to understand and engage with. They proposed nine separate boundaries (see Figure 39). Humanity has already crossed what scientists feel is the acceptable boundary for three of these (see Table 6). These include climate change as reviewed in this book, biodiversity loss which is currently 100–1,000 times the background rate and the disruption of key biogeochemical cycles. For example, the early 20th century invention of the Haber–Bosch process allowing the conversion of atmospheric nitrogen to ammonia for use as fertilizer has altered the global nitrogen cycle so radically that the nearest geological comparison are events ~2.5 billion years ago.

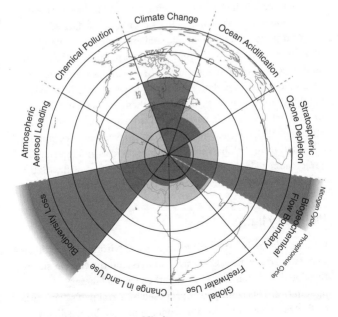

39. Planetary environmental limits

Table 6. Planetary boundaries

Boundary	Control variable	Boundary crossed
1. Climate change	Atmospheric carbon dioxide concentration (ppm by volume)	yes
	Alternatively: Increase in radiative forcing (W/m^2) since the start of the industrial revolution (~1750 CE)	yes
2. Biodiversity loss	Extinction rate (number of species per million per year)	yes
3. Biogeochemical	(a) anthropogenic nitrogen removed from the atmosphere (millions of tonnes per year)	yes
	(b) anthropogenic phosphorus going into the oceans (millions of tonnes per year)	no
4. Ocean acidification	Global mean saturation state of aragonite in surface seawater (omega units)	no
5. Land use	Land surface converted to cropland (per cent)	no
6. Freshwater	Global human consumption of water (km^3/yr)	no
7. Ozone depletion	Stratospheric ozone concentration (Dobson units)	no
8. Atmospheric aerosols	Overall particulate concentration in the atmosphere, on a regional basis	Not defined
9. Chemical pollution	Concentration of toxic substances, plastics, endocrine disruptors, heavy metals, and radioactive contamination into the environment	Not defined

The global phosphorus and sulphur cycles have seen similar dramatic alterations. Of the other six boundaries, we are not yet at the proposed boundary for four of them and scientists are unable as yet to define a boundary at which two of them will become a planetary problem. The latter two include atmospheric aerosols and chemical pollution. The concept of planetary boundaries has come under criticism as it is very hard to define what a safety zone is within each theme and also there is no current way to examine the interaction between the boundaries, for example the how climate change may effect biodiversity. But the concept of planetary boundaries is extremely useful as it challenges the belief that resources are either limitless or infinitely substitutable. It threatens the business-as-usual approach to economic growth. As Professor Will Steffen pointed out, the fact that reference to planetary boundaries was excluded from the Rio+20 Earth Summit conference statement (2012) is, counterintuitively, a sign that the concept is being taken very seriously and has indeed gained enough traction to be threatening to the status quo.

The planetary boundaries concept only deals with the physical environment and there is a strong argument that human societal boundaries should also be included. These could be access to food, water, health services, education, and energy. At the moment, the state of humanity is completely unacceptable: seven million children die needlessly each year; 700 million people go to bed hungry each night; 1,000 million people do not have regular access to clean safe drinking water; 1,100 million people do not have access to electricity; and 3,500 million people live on less than $3.25 per day. The UN Millennium Development Goals (MDGs) have had a significant impact over the last 15 years in reducing poverty and increasing life expectancy, though all agree there is still a very long way to go. Concurrent with the climate change negotiations are UN negotiations on the post-2015 MDGs, which will be called the Sustainable Development Goals that will encapsulate the need for poverty eradication while protecting the

environment. Hence climate change and environment considerations will be central to the UN view of economic development.

Perfect storm

The planetary boundaries concept provides a current view of the state of the global environment. However, with rapid economic development, the environment will come under increasing pressure. For example, by 2030 global food and energy demand will have increased by 50 per cent and water requirement will have increased by 30 per cent. This is partly due to the rise in global population but most is caused by the rapid development of lower income countries and a huge increase in consumption. Add to this the growing effects of climate change, which, as already discussed, directly threaten water and food security, and you have what Sir John Beddington (previous UK government chief scientific adviser) calls the 'perfect storm' (see Figure 40).

We also have to consider the effect of a growing global population. According to UN predictions, global population will increase to nine billion people by 2050, when it will stabilize and may even drop slightly. However, that means there will be a further two billion people on the planet in the next 35 years. This is because as countries develop they go through a demographic transition (see Figure 41). Demographic transition refers to the transition from high birth and death rates to low birth and death rates as a country develops from a pre-industrial to an industrialized economic system. However, infant and mother mortality rates are quick to drop as an area obtains better sanitation and healthcare provision. A societal shift to lower birth rates can take a long time, and it is during this time of transition that a large increase in population occurs. Globally, the highest increase in global population was in the 1960s with growth of over 2 per cent per

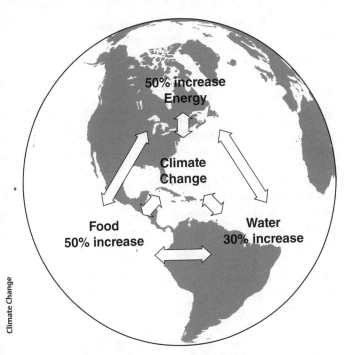

40. Global energy, food, and water demand by 2030

year due to the roll out of mass immunization programmes and the eradication of small pox. The most effective solution to high birth rates has been the education of women up to at least secondary school level, as they then take control of their own fertility. There are huge disparities globally in population growth, with regions like Bihar and Uttar Pradesh in India set to double their population in the next 50 years, while some very developed countries now have shrinking populations. In general the greatest population increases are occurring in the least developed countries. It is this growth in population in the poorest and most vulnerable countries that will make the impacts of climate change worse.

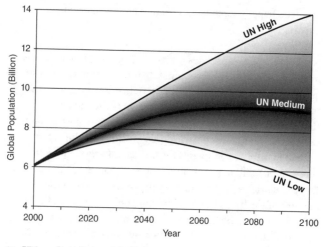

41. UN predicted growth in global population growth

Neoliberal myth

Scientists struggle with the fact that despite the huge weight of
evidence, a vocal but significant minority of developed countries
refuse to accept that climate change is happening. Their response
is to go out and collect even more evidence against it. This is
called the 'deficit model', when scientist perceive that decisions are
not being taken due to a lack of information. But social scientists
have found that acceptance of climate change as a real
phenomenon has little to do with science and everything to do
with politics. Climate change challenges the Anglo-American
neoliberal view that is held so dear by mainstream economists and
politicians. Climate change shows a fundamental failure of the
market, and requires governments to act collectively to regulate
industry and business. It is one of the greatest ironies of climate
change that the very politicians who are denying climate change
because of free market values are also the ones that are happiest to
endorse over $500 billion of subsides for the fossil fuel industry
per year. It is a myth that there is any truly free market, with many

countries happily supporting subsidies and the blocking of imports.

Neoliberalism encapsulates a set of beliefs which include: the need for markets to be free; for state intervention to be as small as possible; strong private property rights; low taxation; and individualism. Underlying neoliberalism is the seductive view that it provides a market-based solution that enables everyone to become wealthier. This trickle-down effect has been the central mantra of neoliberals for the last 35 years. Currently there are 3.5 billion people living on less than $3.25 per day. In fact the 85 richest people in the world currently own the same amount of wealth as those 3.5 billion poorest people collectively. If we want to eradicate poverty and bring the very poorest people in the world up to $3.25 per day then according to David Woodward at current rates of trickle-down it would require global GDP to increase by 15 times, taking over 100 years. If we want to be more radical and lift everyone in poverty to $5 per day then it would require global GDP to increase by 170 times, taking over 200 years. Under the current economic system, huge consumption levels would be required in the developed and rapidly developing countries for world GDP to rise by 15 times, let alone 170. All of this would lead to great energy and resource demands, and hence higher carbon emissions and accelerated climate change. So the neoliberal nightmare is that to lift people out of poverty there is a perception that we need make and consume more stuff. This all requires cheap energy, which will mainly come from fossil fuels, and more land for agriculture, driving deforestation and detrimental land use changes. This will ultimately lead to accelerated climate change, making those poorest of people even more vulnerable to extreme weather events. So fundamentally, climate change challenges the very economic theories that have dominated global economics for the last 35 years. Hence is it any wonder that many people prefer climate change denial to having to face the prospect of building a new political system that will

allow collective action and more equal distribution of wealth, resources, and opportunities?

Taking action

With such clear evidence that climate change and global environmental degradation is occurring and is going to get worse, why has so very little action been taken? The first excuse used by politicians is that there is scientific uncertainty, which means it is not possible to act. In the face of scientific uncertainty, various philosophies for decision-making have arisen. Each has flaws. The precautionary principle says that action should be taken against worst-case scenarios, 'just in case'. This is problematic as it does not take into account that acting, as well as not acting, may yield unacceptable consequences. Moreover, the precautionary principle as outlined in both Agenda 21 and post-Rio discussion only looks at 'cost-effective measures' when it comes to assessing uncertainty. In effect, this means any risk that is not fully quantifiable due to uncertainty can be ignored. So politically the precautionary principle has been reduced to a type of cost-benefit analysis.

Cost-benefit analyses attempt to take into account the full cost of a risk by totalling the sum impacts of different actions or non-actions. But there are serious problems in accounting for all possible costs and giving them a numerical value. Debates arise about whether the cost to our offspring should count for less than the costs to today's generation (known as 'discounting the future'), and about the (perhaps variable) value of human life. Can we morally argue that a life of an Indian is worth less than that of an American? Economists' models do make this assumption. Cost-benefit analysis and mainstream economics also fail to account for natural capital and ecosystem services. Natural capital is the stock of natural ecosystems that yields a flow of valuable ecosystem goods or services into the future. For example, a stock of trees or fish provides a flow of new trees or fish, a flow that can

be indefinitely sustainable. While ecosystem services are services provided for nature that are essential for human life, they are split into four main categories: *provisioning*, such as the production of food and water; *regulating*, such as the control of climate and disease; *supporting*, such as nutrient cycles and crop pollination; and *cultural*, such as spiritual and recreational benefits. It is clear that many of these ecosystem services are of huge economic value. Another economic approach has been the calculation of the social cost of carbon (SCC). SCC is the calculation of the societal cost of emitting an additional tonne of GHGs into the atmosphere. However, the SCC is heavily influenced by what is defined as climate change damage, how risk averse society should be, and how much future impacts can discounted due to increased wealth. Published estimates of the SCC range from $6 to $445 per tonne of carbon dioxide. However, a recent paper by van den Bergh and Botzen in *Nature Climate Change* suggests the lower limit should be $125 per tonne of carbon dioxide. Hence the *RCP 8.5* suggests that by 2050 we will emitting at least another half a trillion tonnes which will cost the global economy a minimum of $60 trillion to deal with.

While uncertainty rarely stops politicians making decisions, in the case of climate change public opinion and scientific uncertainty nonetheless seem to be used as excuses for inaction. For example, politicians used to say 'we need to wait until scientists prove that mankind is causing climate change'. That hurdle has passed, now they have moved on to 'we need to wait until scientists can tell us exactly what will happen, and the costs of inaction', or, 'we need to wait for public opinion to be behind action'. As has been demonstrated in Chapters 3 and 4, the former will never occur because the modelling can never provide an absolute level of certainty. The latter is a sleight of hand as in no other area do politicians assume the need for public support on major national decisions. For example, look at decision-making on matters ranging from wars to bank bailouts, from taxation to healthcare reforms. Moreover governments are extremely risk-averse. Hence

it is clear that politicians see inaction as a safer option than making new policies that may fail; politicians put more value on avoiding blame than they do on gaining credit.

Win-win solutions

With multiple global challenges in the 21st century, action on climate change should always contain an element of win-win. For example, supporting a huge increase in renewable energy not only reduces emissions but helps to provide a country with energy security by reducing the reliance on imported oil, coal, and gas. Reduced deforestation and increased reforestation should not only draw down additional carbon dioxide from the atmosphere but it will help to retain biodiversity, stabilize soils and local rainfall, and provide livelihoods for local people via carbon credits. Measures that reduce car use will increase walking and cycling, which in turn will reduce obesity and heart attacks. Ensuring that women are educated to at least secondary school level all around the world will empower them to take control of their own fertility, and this in turn will help to stabilize population growth and pressures on development. No one can object to creating a better world, even if we are extremely lucky and the scale of climate change is at the low end of all projections. This point is beautifully illustrated by a cartoon drawn by Joel Prett that was first published in *USA Today* in 2009, concerning the Copenhagen Conference (see Figure 42). What climate change does do is to challenge our view of the future and show us that more of the same will not work. What is required is a new vision of our world and how people can have full access to fundamental rights such as: clean air and water, a nutritionally balanced diet, suitable housing, free healthcare, free education, and full employment.

Conclusions

Climate change is one of the few areas of science that makes us examine the whole basis of modern society. It is a subject that has

42. *USA Today* cartoon of the Copenhagen climate conference

politicians arguing, sets nations against each other, queries individual choices of lifestyle, and ultimately asks questions about humanity's relationship with the rest of the planet. There is very little doubt that climate change will accelerate in this century; our best estimates suggest a global mean surface temperature rise of between 2.8°C and 5.4°C by the end of the 21st century. Sea level is projected to rise by between 52 cm and 98 cm by 2100 with significant changes in weather patterns, and more extreme climate events. This is not the end of the world as was envisaged by many environmentalists in the late 1980s and early 1990s, but it does mean a huge rise in misery for billions of people.

Climate change is the major challenge for our global society. We should not underestimate the work that ahead of us in this regard. Despite 30 years of climate change negotiations there has been no deviation in emissions from the business-as-usual pathway. With the rapid development of the BASIC (Brazil, South Africa, India, and China) countries there is little hope that emissions will be curbed in the near future. Moreover, the

consensus approach used by the IPCC to secure agreement from all parties means it is inherently conservative. We should perhaps therefore view the top estimates of climate change as those more likely to occur. This means we are staring down the barrel of a gun, with warming of over 5°C very likely by 2100, if we do not start to act now. Add to this the estimates of top economists that it could cost us over 20 per cent of everything the world earns in the future to deal with a warmer world. This contrasts with a figure of only 2–3 per cent of what we currently earn to convert our global economy to low carbon. Even if the cost-benefits are not so great from an economics point of view, the ethical case for paying now to prevent the deaths of tens of millions of people and to avoid the likely increase in human misery must be clear.

So what are the solutions to climate change? First, there must be an international political solution; without a post-2015 agreement we are looking at huge increases in global carbon emissions and devastating climate change. Any political agreement will have to include plans to protect the rapid development of developing countries, as it is a moral imperative that people in the poorest countries have the right to develop and to obtain a similar level of healthcare, education, and life expectancy to that in the West. Climate change policies and laws based around international negotiations must be implemented at both regional and national levels to provide multi-levels of governance that will ensure these cuts in emissions really do occur. Novel ways of redistributing wealth, globally as well as within nations, are needed to lift billions of people out of poverty without huge increases in consumption and resource depletion. Support and money is also needed to help developing countries to adapt to the climate changes that will inevitably happen.

Second, we must greatly increase the funding for developing cheap and clean energy production, as all economic development is based on increasing energy usage. Fossil fuel subsidies should

be made illegal and sanctions applied to countries that continue to skew the world's energy markets. To tackle climate change we really need the level of funding that is usually only ever achieved when a country is at war. For example USA has spent between $4 and $6 trillion on the war in Iraq and Afghanistan; just imagine if all that money had been put into developing technology for a zero-carbon world. The International Energy Agency estimates $20 trillion will be invested in energy over the next 15 years—what we must do is to ensure that it is not in fossil fuels. But even if renewable energy technology does become available, there is no guarantee that it would be made affordable to all nations, since we live in a world where even life-saving drugs are costed in a way that will achieve maximum profit. Nor is there any guarantee that if we had unlimited free energy it would prevent us from continuing to abuse the planet. Professor Paul Ehrlich at Stanford University, commenting on the possibility of unlimited clean energy from cold fusion, suggested it would be 'like giving a machine gun to an idiot child'.

We must not pin all our hopes on global politics and clean energy technology, so we must prepare for the worst and adapt. If implemented now, a lot of the costs and damage that could be caused by changing climate can be mitigated. This requires nations and regions to plan for the next 50 years, something that most societies are unable to do because of the very short-term nature of politics. So climate change challenges the very way we organize our society. Not only does it challenge the concept of the nation-state versus global responsibility, but also the short-term vision of our political leaders. To answer the question of what we can do about climate change, we must change some of the basic rules of our society to allow us to adopt a much more global and long-term approach. In doing so we may also provide solutions to global environmental degradation, poverty, and security.

Further reading

History of climate change

Corfee-Morlot, J., et al. Climate Science in the Public Sphere, *Philosophical Transactions A of the Royal Society*, 365/1860 (2007): 2741–76.

Leggett, J. K. *Half-Gone: Oil, Gas, Hot Air and the Global Energy Crisis* (Portobello, 2006).

Lovelock, J., *The Ages of Gaia* (Norton, 1995).

Mann, M. *The Hockey Stick and the Climate Wars: Dispatches from the Front Lines* (Columbia University Press, 2013).

Oreskes, N. and M. Conway *Merchants of Doubt: How a Handful of Scientists Obscured the Truth on Issues from Tobacco Smoke to Global Warming* (Bloomsbury Paperbacks, 2012).

Weart, S. R. *The Discovery of Global Warming, New Histories of Science, Technology, and Medicine* (Harvard University Press, 2003).

Science

Archer, D. *Global Warming: Understanding the Forecast*, 2nd edn (John Wiley & Sons, 2011).

Dessler, A. E. *Introduction to Modern Climate Change* (Cambridge University Press, 2012).

Houghton, J. T. *Global Warming: The Complete Briefing*, 4th edn (Cambridge University Press, 2009).

IPCC, Climate Change 2013—The Physical Science Basis Contribution of Working Group I to the Fifth Assessment Report of the

Intergovernmental Panel on Climate Change (Cambridge
University Press, 2014).

Maslin, M. A. and S. Randalls (eds) *Routledge Major Work Collection.
Future Climate Change: Critical Concepts in the Environment*
(4 Volumes containing reproductions of 85 of the most important
papers published on climate change) (Routledge, 2012).

Ruddiman, W. F. *Earth's Climate: Past and Future*, 2nd edn
(W. H. Freeman, 2007).

Impacts

Costello, A., et al. Managing the Health Effects of Climate Change, *The
Lancet*, 373 (2009): 1693–733.

Garcia R.A., et al. Muliple Dimensions of Climate Change and their
Implications for Biodiversity, *Science*, 344 (2014): 486–96.

IPCC, Climate Change 2014—Impacts, Adaptation, and Vulnerability,
Contribution of Working Group II to the Fifth Assessment Report
of the Intergovernmental Panel on Climate Change (Cambridge
University Press, 2014).

National Climate Assessment. U.S. Global Change Research Program
(2014) <http://nca2014.globalchange.gov/report>.

Stern, N. *The Economics of Climate Change: The Stern Review*
(Cambridge University Press, 2007)

Politics and governance

Giddens, A. *The Politics of Climate Change*, 2nd edn (Polity Press,
2011).

Gupta, J. *The History of Global Climate Governance* (Cambridge
University Press, 2014).

IPCC, Climate Change 2014—Mitigation of Climate Change,
Contribution of Working Group III to the Fifth Assessment Report
of the Intergovernmental Panel on Climate Change (Cambridge
University Press, 2014).

Labatt S. and R. R. White *Carbon Finance* (Wiley, 2007).

Meyer, A. *Contraction and Convergence: The Global Solution to
Climate Change* (Green Books, 2000).

Schellnhuber, H. J., et al. *Avoiding Dangerous Climate Change*
(Cambridge University Press, 2006).

Solutions

Centre for Alternative Technology, *Zero Carbon Britian 2030: A New Energy Strategy* (CAT Publications, 2010)

Grubb, M. *Planetary Economics: Energy, Climate Change and the Three Domains of Sustainable Development* (Earthscan from Routledge, 2014)

Hamilton, C. *Earthmasters: The Dawn of the Age of Climate Engineering* (Yale University Press, 2014).

Helm, D. *The Carbon Crunch: How We're Getting Climate Change Wrong—and How to Fix it* (Yale University Press, 2013).

IPCC, Climate Change 2014—Impacts, Adaptation, and Vulnerability, Contribution of Working Group II to the Fifth Assessment Report of the Intergovernmental Panel on Climate Change (Cambridge University Press, 2014).

Jackson, T. *Prosperity without Growth: Economics for a Finite Planet* (Earthscan from Routledge, 2009).

Roaf, S., et al. *Adapting Building and Cities for Climate Change* (Elsevier, 2005).

Royal Society, Geoengineering the Climate: Science, Governance and Uncertainty, The Royal Society Science Policy Centre Report, The Royal Society, 10/09 (2009): 81.

Walker, G., and D. King *The Hot Topic* (Bloombury, 2008)

General reading

Berners-Lee, M., and J. Clark *The Burning Question: We Can't Burn Half the World's Oil, Coal and Gas. So How Do We Quit?* (Profile Books, 2013).

Hansen, J. *Storms of My Grandchildren: The Truth about the Coming Climate Catastrophe and Our Last Chance to Save Humanity* (Bloomsbury Paperbacks, 2011).

Henson, R. *The Rough Guide to Climate Change*, 3rd edn (Rough Guides, 2011).

Hulme, M. *Why We Disagree About Climate Change: Understanding Controversy, Inaction and Opportunity* (Cambridge University Press, 2009).

Lomborg, B. *The Skeptical Environmentalist: Measuring the Real State of the World* (Cambridge University Press, 2001).

Lynas, M. *Six Degrees: Our Future on a Hotter Planet* (Fourth Estate, 2007).

Maslin, M. *Climate: A Very Short Introduction* (Oxford University Press, 2013).

Royal Society, People and the Planet, The Royal Society Science Policy Centre Report 01/12 (2012): 81.

Sachs, J. *The End of Poverty* (Penguin, 2005).

Tickell, O. *Kyoto2: How to Manage the Global Greenhouse* (Zed Books, 2008).

Fiction inspired by climate change

Astley N. (ed.) *Earth Shattering: Ecopoems* (Bloodaxe Books, 2007).

Cowley J. (ed.) *Granta 102: The New Nature Writing* (Granta, 2008).

Evans, K. *Funny Weather* (Myriad Editions, 2006).

Griffiths, J. *WILD—An Elemental Journey* (Penguin Books, 2008).

Hamilton, P. F. *Mindstar Rising* (Pan Books, 1993).

McEwan, I. *Solar* (Vintage, 2011).

McNeil, J. *The Ice Lovers—A Novel* (McArthur & Company, 2009).

Robinson, K. S. *Forty Signs of Rain* (HarperCollins, 2004).

Winterson, J. *The Stone Gods* (Hamish Hamilton, 2007).

Index

Index

ONLINE CATALOGUE
A Very Short Introduction

Our online catalogue is designed to make it easy to find your ideal Very Short Introduction. View the entire collection by subject area, watch author videos, read sample chapters, and download reading guides.

http://fds.oup.com/www.oup.co.uk/general/vsi/index.html

VERY SHORT INTRODUCTIONS are for anyone wanting a stimulating and accessible way in to a new subject. They are written by experts, and have been translated into more than 40 different languages.

The Series began in 1995, and now covers a wide variety of topics in every discipline. The VSI library now contains over 350 volumes—a Very Short Introduction to everything from Psychology and Philosophy of Science to American History and Relativity—and continues to grow in every subject area.

Very Short Introductions available now:

Climate Change: A Very Short Introduction